THE BOTANIC GARDEN BY
ERASMUS DARWIN

THE PICKERING MASTERS SERIES

THE BOTANIC GARDEN BY ERASMUS DARWIN

Edited by
Adam Komisaruk and Allison Dushane

Volume II
The Loves of the Plants

Routledge
Taylor & Francis Group
LONDON AND NEW YORK

First published 2017
by Routledge
4 Park Square, Milton Park, Abingdon, Oxon OX14 4RN

and by Routledge
605 Third Avenue, New York, NY 10017

Routledge is an imprint of the Taylor & Francis Group, an informa business

British Library Cataloguing-in-Publication Data
A catalogue record for this book is available from the British Library

Library of Congress Cataloging-in-Publication Data
Names: Darwin, Erasmus, 1731-1802, author.
Title: The botanic garden by / Erasmus Darwin.
Description: Milton Park, Abingdon, Oxon ; New York : Routledge, 2017. |
Series: The Pickering Masters series | Includes bibliographical references.
Identifiers: LCCN 2016054656 (print) | LCCN 2017014344 (ebook) |
ISBN 9781315534695 (set) | ISBN 9781315534657 (1) |
ISBN 9781315534534 (2) | ISBN 9781848935655 (set) |
ISBN 9781138691575 (1) | ISBN 9781138691605 (2)
Subjects: LCSH: Botany—Poetry. | Plants—Poetry.
Classification: LCC PR3396 (ebook) | LCC PR3396 .A7 2017 (print) |
DDC 821/.6—dc23
LC record available at https://lccn.loc.gov/2016054656

ISBN: 978-1-8489-3565-5 (Set)
ISBN: 978-1-138-69160-5 (Volume II) **(hbk)**
eISBN: 978-1-315-53453-4 (Volume II) (ebk)

Typeset in Times New Roman
by Apex CoVantage, LLC

Publisher's Note

References within each chapter are as they appear in the original complete work

This project is presented with financial assistance from the West Virginia Humanities Council, a state affiliate of the National Endowment for the Humanities. Any views, findings, conclusions or recommendations do not necessarily represent those of the West Virginia Humanities Council or the National Endowment for the Humanities.

Generous subvention was provided by Furthermore: a program of the J.M. Kaplan Fund.

CONTENTS

ABBREVIATIONS

Amoen. Acad.	*Amoenitates academicae* (Stockholm and Leipzig)
AN	Additional Notes to *The Economy of Vegetation*
BG	*The Botanic Garden*
CL	*Collected Letters of Samuel Taylor Coleridge*, ed. Earl Leslie Griggs, 6 vols. (Oxford: Clarendon P, 2002)
CW	*The Collected Works of Samuel Taylor Coleridge*, ed. James Engell et al., 16 vols. (Princeton: Princeton UP, 1969–2001)
EV	*The Economy of Vegetation* (prepended number indicates the edition, appended number the year of publication; e.g., *3EV1795* = third edition of *EV*, 1795)
Edinb. Trans.	*Transactions of the Royal Society of Edinburgh*
Exper. Air	Joseph Priestley, *Experiments and Observations on Different Kinds of Air*, 3 vols. (London, 1774–77)
Exper. Nat. Phil.	Joseph Priestley, *Experiments and Observations Relating to Various Branches of Natural Philosophy*, 3 vols. (London and Birmingham, 1779–86)
Exper. Elec.	Benjamin Franklin, *Experiments and Observations on Electricity, Made at Philadelphia in America* (London, 1751)
Hist. de l'Acad. roy.	*Histoire de l'Académie royale des sciences* (Paris)
Kön. Akad. Abh.	*Königlich-Schwedischen Akademie der Wissenschaften Abhandlungen* (Hamburg and Leipzig)
Kongl. Vet.-Acad. Handl.	*Kongliga [Svenska] Vetenskaps-akademiens Handligar* (Stockholm)
Letters	*The Collected Letters of Erasmus Darwin*, ed. Desmond King-Hele (New York and Cambridge, 2007)
LP	*The Loves of the Plants* (prepended number indicates the edition, appended number the year of publication; e.g., *5LP1799* = fifth edition of *LP*, 1799)

Phil. Trans.	*Philosophical Transactions of the Royal Society* (London)
PW	*The Poetical Works of Anna Seward*, ed. Walter Scott, 3 vols. (London, 1810)
SV	*A System of Vegetables*, trans. Erasmus Darwin after Linnaeus (Lichfield, 1783–85)

ILLUSTRATION CREDITS

ILLUSTRATIONS

Plate 1 The British Library Board, Shelfmark [illegible] Part [illegible]
opp. p. [illegible]
Plate 2 The British Library Board, Shelfmark [illegible] Part [illegible]
opp. p. [illegible]
Plate 3 The British Library Board, Shelfmark [illegible] Part [illegible]
opp. p. [illegible]
Plate 4 The British Library Board, Shelfmark [illegible] Part [illegible]
opp. p. [illegible]
Plate 5 The British Library Board, Shelfmark [illegible] Part [illegible]
opp. p. [illegible]
Plate 6 The British Library Board, Shelfmark [illegible] Part [illegible]
opp. p. [illegible]
Plate 7 The British Library Board, Shelfmark [illegible] Part [illegible]
opp. p. [illegible]
Plate 8 The British Library Board, Shelfmark [illegible] Part [illegible]
opp. p. [illegible]
Plate 9 The British Library Board, Shelfmark [illegible] Part [illegible]
opp. p. [illegible]
Plate 10 The British Library Board, Shelfmark [illegible] Part [illegible]
opp. p. [illegible]
Plate 11 The British Library Board, Shelfmark [illegible] Part [illegible]
opp. p. [illegible]
Plate 12 The British Library Board, Shelfmark [illegible] Part [illegible]
opp. p. [illegible]
Plate 13 The British Library Board, Shelfmark [illegible] Part [illegible]
opp. p. [illegible]

EDITORIAL PRINCIPLES

This eclectic edition of *The Botanic Garden* pairs the first lifetime editions of the respective parts (*1EV1792* and *1LP1789*). Transcriptions were made from the British Library copies digitized in *Eighteenth-Century Collections Online*; and checked against physical copies held in the Folger Shakespeare Library in Washington, DC, the Library of Congress, and the Harry Ransom Center at the University of Texas, Austin. A clear reading text is presented, with apparatus limited to Darwin's own (unnumbered) footnotes. The placement and text of the Advertisement corresponds to *1EV1792* (the first time *EV* was paired with *LP*) rather than to *1LP1789* (the first time the Advertisement appeared). Material appended to the first editions has been incorporated into the main text where Darwin's intentions to this effect can be inferred (e.g., from the heading "Verses omitted by mistake"). Especially long Additional Notes and Supplements to *1LP1789* have been left as back matter, however, as they were by Darwin in subsequent editions.

In most cases, transcriptions defer to the first editions. Idiosyncrasies of spelling and punctuation have been retained unless (1) they interfere with intelligibility (e.g., "car'd" for "carr'd"), in which case they have been emended and noted; (2) they are included in Darwin's own errata to the first editions, in which case they have been emended and noted as such; or (3) they are likely due to printing errors (e.g., "f" for long "s"), in which case they have been emended silently. Darwin's error-ridden line numbers have been regularized, and his references thereto emended silently.

Explanatory notes reference poetry by canto and line number, Darwin's footnotes by poetry-line number with an "n", and prose by section and paragraph number. *AN*, because of its length, receives its own sections of the apparatus but is otherwise treated as integral to *EV*. Items repeated across *EV* and *LP* are annotated uniquely for each work. Items repeated within each *EV* and *LP* are fully annotated at the first occurrence, cross-referenced at the second occurrence (or, for plants in *LP*, the major occurrence) and unannotated thereafter. Items repeated across *EV* and *AN* are cross-referenced to the first occurrence in *EV* at the first occurrence in *AN*, and unannotated thereafter. Unique items repeated within *AN* are fully annotated at the first occurrence, cross-referenced at the second occurrence and unannotated thereafter.

References to minerals in *EV* and *AN* are presented in the separate Table of Minerals. The taxonomy is that used in Richard Kirwan, *Elements of Mineralogy* (London, 1784). Each mineral is listed by the canto and line number (or Additional Note and paragraph number) of the first reference; followed by name, class, genus, species, and variety or series if applicable.

References to plant genera in *LP* are presented in the separate Table of Plants. The taxonomy is that used in *SV*. Each genus is listed by the canto and line number of the first substantive reference; the canto and line number of the major reference if different from the previous; the poetic name given in *LP* if applicable; common English name(s) as given in *LP* and *SV* (with corresponding Linnaean species in parentheses); Linnaean class, order, and genus or family; and present-day order, family and genus given by the 2009 classification system of the Angiosperm Phylogeny Group (APG III). Equivalences were verified using the database of the Linnaean Plant Name Typification Project, Natural History Museum, London.

The four English lifetime editions of *EV* and the five English lifetime editions of *LP* are collated. Textual and most other apparatus refers to editions of *EV* and *LP* independently, rather than jointly as *BG*; for the rationale, see "Text" section of Editors' Introduction. All editions consulted were British Library copies digitized in *ECCO* and checked against Folger, Library of Congress and Ransom copies. Textual notes record only deviations from the first editions. A variant is noted at its first occurrence only, with "*ff.*" indicating that it obtains throughout (e.g., "**1.115n *ff.* lightening]** lightning *4EV1799*"); only exceptions are noted subsequently. Variants in punctuation, capitalization and line numbering are not noted. A plus sign indicates that the variant appears in all subsequent editions (e.g., *3LP1791+* for *3LP1791*, *4LP1794* and *5LP1799*). Sub-references to inserted passages use the number of the line or paragraph immediately preceding the insertion, an "a" and the number of the line within the insertion (e.g., "1.358a3" for the third line of the passage inserted after 1.358).

Illustrations are reproduced from the first editions in which they appeared, except for the four from *1LP1789* that Darwin clearly found unsatisfactory (LP2–5), in which case *2LP1790* is used. Plates not intended for the first editions (EV4a, AN4a, LP7a-b, LP8a-b) are removed to the Additional Illustrations sections, but numbered in such a way as approximately to indicate their placement upon first publication (e.g., LP8b fell between LP8 and LP9 in *2LP1790*, but LP8a fell between LP8 and LP8b in *5LP1799*). In the inscriptions and imprints, a single slash (|) signifies a division between lines, a double slash (||) a division within a line.

LP1

THE BOTANIC GARDEN, PART II. CONTAINING

The LOVES of the PLANTS,

A POEM.

WITH PHILOSOPHICAL NOTES.

VOLUME THE SECOND.

Vivunt in Venerem frondes; nemus omne per altum
Felix arbor amat; nutant ad mutua Palmæ
Fædera, populeo suspirat Populus ictu,
Et Platani Platanis, Alnoque assibilat Alnus.

Claud. epith.

LICHFIELD: Printed by *J. Jackson.*
Sold by J. Johnson, St. Paul's Church Yard, London:
M, DCC, LXXXIX.
[Entered at Stationer's Hall.]

PROEM.

GENTLE READER! 1

Lo, here a CAMERA OBSCURA is presented to thy view, in which are lights and 2
shades dancing on a whited canvas, and magnified into apparent life!—if thou art
perfectly at leasure for such trivial amusement, walk in, and view the wonders of
my INCHANTED GARDEN.

Whereas P. OVIDIUS NASO, a great Necromancer in the famous Court of AUGUS- 3
TUS CÆSAR, did by art poetic transmute Men, Women, and even Gods and God-
desses, into Trees and Flowers; I have undertaken by similar art to restore some of
them to their original animality, after having remained prisoners so long in their
respective vegetable mansions; and have here exhibited them before thee. Which
thou may'st contemplate as diverse little pictures suspended over the chimney of
a Lady's dressing-room, *connected only by a slight festoon of ribbons*. And which,
though thou may'st not be acquainted with the originals, may amuse thee by the
beauty of their persons, their graceful attitudes, or the brilliancy of their dress.

FAREWELL. 4

LP2

PREFACE.

Linneus has divided the vegetable world into 24 Classes; these Classes into about 1
120 Orders; these Orders contain about 2000 Families, or Genera; and these Families about 20,000 Species; besides the innumerable Varieties, which the accidents of climate, or cultivation have added to these Species.

The Classes are distinguished from each other in this ingenious system, by the 2
number, situation, adhesion, or reciprocal proportion of the males in each flower. The Orders, in many of these Classes, are distinguished by the number, or other circumstances of the females. The Families, or Genera, are characterized by the analogy of all the parts of the flower or fructification. The Species are distinguished by the foliage of the plant; and the Varieties by any accidental circumstance of colour, taste, or odour; the seeds of these do not always produce plants similar to the parent; as in our numerous fruit-trees and garden flowers; which are propagated by grafts or layers.

The first eleven Classes include the plants, in whose flowers both the sexes 3
reside; and in which the Males or Stamens are neither united, nor unequal in height when at maturity; and are therefore distinguished from each other simply by the number of males in each flower, as is seen in the annexed Plate, copied from the Dictionaire Botanique of M. Bulliard, in which the numbers of each division refer to the Botanic Classes.

CLASS I. One Male, *Monandria*; includes the plants which possess but One 4
Stamen in each flower. 5

II. Two Males, *Diandria*. Two Stamens. 6

III. Three Males, *Triandria*. Three Stamens. 7

IV. Four Males, *Tetrandria*. Four Stamens. 8

V. Five Males, *Pentandria*. Five Stamens. 9

VI. Six Males, *Hexandria*. Six Stamens. 10

VII. Seven Males, *Heptandria*. Seven Stamens. 11

VIII. Eight Males, *Octandria*. Eight Stamens. 12

IX. Nine Males, *Enneandria*. Nine Stamens. 13

X. Ten Males, *Decandria*. Ten Stamens. 14

XI. Twelve Males, *Dodecandria*. Twelve Stamens. 15

The next two Classes are distinguished not only by the number of equal and disunited males, as in the above eleven Classes, but require an additional circumstance
to be attended to, *viz.* whether the males or stamens be situated on the calyx, or not. 16

XII. Twenty Males, *Icosandria*. Twenty Stamens inserted on the calyx or
flower-cup; as is well seen in the last Figure of No. xii. in the annexed Plate. 17

XIII. Many Males, *Polyandria*. From 20 to 100 Stamens, which do not adhere
to the calyx; as is well seen in the first Figure of No. xiii. in the annexed Plate. 18

In the next two Classes, not only the number of stamens are to be observed, but 19
the reciprocal proportions in respect to height.

XIV. Two Powers, *Didynamia*. Four Stamens, of which two are lower than the 20
other two; as is seen in the two first Figures of No. xiv.

XV. Four Powers, *Tetradynamia*. Six Stamens; of which four are taller, and the 21
two lower ones opposite to each other; as is seen in the third Figure of the upper row in No. xv.

The five subsequent Classes are distinguished not by the number of the males, 22
or stamens, but by their union or adhesion, either by their anthers, or filaments, or to the female or pistil.

XVI. One Brotherhood, *Monadelphia*. Many Stamens united by the filaments 23
into one company; as in the second Figure below of No. xvi.

XVII. Two Brotherhoods, *Diadelphia*. Many Stamens united by their fila- 24
ments into two companies; as in the uppermost Fig. No. xvii.

XVIII. Many Brotherhoods, *Polyadelphia*. Many Stamens united by their fila- 25
ments into three or more companies, as in No. xviii.

XIX. Confederate Males, *Syngenesia*. Many Stamens united by their anthers; 26
as in first and second Figures, No. xix.

XX. Feminine Males, *Gynandria*. Many Stamens attached to the pistil. 27

The next three Classes consist of plants, whose flowers contain but one of the 28
sexes; or if some of them contain both sexes, there are other flowers accompanying them of but one sex.

XXI. One House, *Monœcia*. Male flowers and female flowers separate, but on 29
the same plant.

XXII. Two Houses, *Diœcia*. Male flowers and female flowers separate, on dif- 30
ferent plants.

XXIII. Polygamy, *Polygamia*. Male and female flowers on one or more plants, 31
which have at the same time flowers of both sexes.

The last Class contains the plants whose flowers are not discernible. 32

XXIV. Clandestine Marriage, *Cryptogamia*. 33

The Orders of the first thirteen Classes are founded on the number of Females, 34
or Pistils, and distinguished by the names, One Female, *Monogynia*. Two Females, *Digynia*. Three Females, *Trigynia*. *&c.* as is seen in No. 1. which represents a plant of one male, one female; and in the first Figure of No. xi. which represents a flower with twelve males, and three females; (for, where the pistils have no apparent styles, the summits, or stigmas, are to be numbered) and in the first Figure of No. xii. which represents a flower with twenty males and many females; and in the last Figure of the same No. which has twenty males and one female; and in No. xiii. which represents a flower with many males and many females.

35 The class of Two Powers, is divided into two natural Orders; into such as have their seeds naked at the bottom of the calyx, or flower-cup; and such as have their seeds cover'd; as is seen in No. xiv. Fig. 3. and 5.

36 The class of Four Powers, is divided also into two Orders; in one of these the seeds are inclosed in a silicule, as in *Shepherd's purse*. No. xiv. Fig. 5. In the other they are inclosed in a silique, as in *Wall-flower*. Fig. 4.

37 In all the other Classes, excepting the Classes Confederate Males, and Clandestine Marriage, as the character of each Class is distinguished by the situations of the males; the character of the Orders is mark'd by the numbers of them. In the Class One Brotherhood, No. xvi. Fig. 3. the Order of ten males is represented. And in the Class Two Brotherhoods, No. xvii. Fig. 2. the Order ten males is represented.

38 In the Class Confederate Males, the Orders are chiefly distinguished by the fertility or barrenness of the florets of the disk, or ray of the compound flower.

And in the Class of Clandestine Marriage, the four Orders are term'd Ferns, Mosses, Flags, and Fungusses.

39 The Orders are again divided into Genera, or Families, which are all natural associations, and are described from the general resemblances of the parts of fructification, in respect to their number, form, situation, and reciprocal proportion. These are the Calyx, or Flower-cup, as seen in No. iv. Fig. 1. No. x. Fig. 1. and 3. No. xiv. Fig. 1. 2. 3. 4. Second, the Corol, or Blossom, as seen in No. i. ii. &c. Third, the Males, or Stamens; as in No. iv. Fig. 1. and No. viii. Fig. 1. Fourth, the Females, or Pistils; as in No. i. No. xii. Fig. 1. No. xiv. Fig. 3. No. xv. Fig. 3. Fifth, the Pericarp or Fruit-vessel; as No. xv. Fig. 4. 5. No. xvii. Fig. 2. Sixth, the Seeds.

40 The illustrious author of the Sexual System of Botany, in his preface to his account of the Natural Orders, ingeniously imagines, that one plant of each Natural Order was created in the beginning; and that the intermarriages of these produced one plant of every Genus, or Family; and that the intermarriages of these Generic, or Family plants produced all the Species: and lastly, that the intermarriages of the individuals of the Species produced the Varieties.

41 In the following Poem, the name or number of the Class, or Order of each plant is printed in italics; as "*Two* brother swains." "*One* House contains them." and the word "*secret.*" expresses the Class of Clandestine Marriage.

42 The Reader, who wishes to become further acquainted with this delightful field of science, is advised to study the works of the Great Master, and is apprized that they are exactly and literally translated into English, by a Society at Lichfield, in four Volumes Octavo.

43 To the SYSTEM OF VEGETABLES is prefix'd a copious explanation of all the Terms used in Botany, translated from a thesis of Dr. Elmsgreen, with the plates and references from the Philosophia Botanica of Linneus.

To the FAMILIES OF PLANTS is prefix'd a Catalogue of the names of plants, 44
and other Botanic Terms, carefully accented, to shew their proper pronunciation; a work of great labour, and which was much wanted, not only by beginners, but by proficients in Botany.

The SYSTEM OF VEGETABLES translated from the Systema Vegetabilium, in 45
two Vols. is sold by Leigh and Sotheby, *York Street, Covent-Garden*: Price 18 Shillings, in Boards.

The FAMILIES OF PLANTS translated from the Genera Plantarum, in 46
two Vols. by Johnson, *St. Paul's Church-Yard*, London: Price 16 Shillings, in Boards.

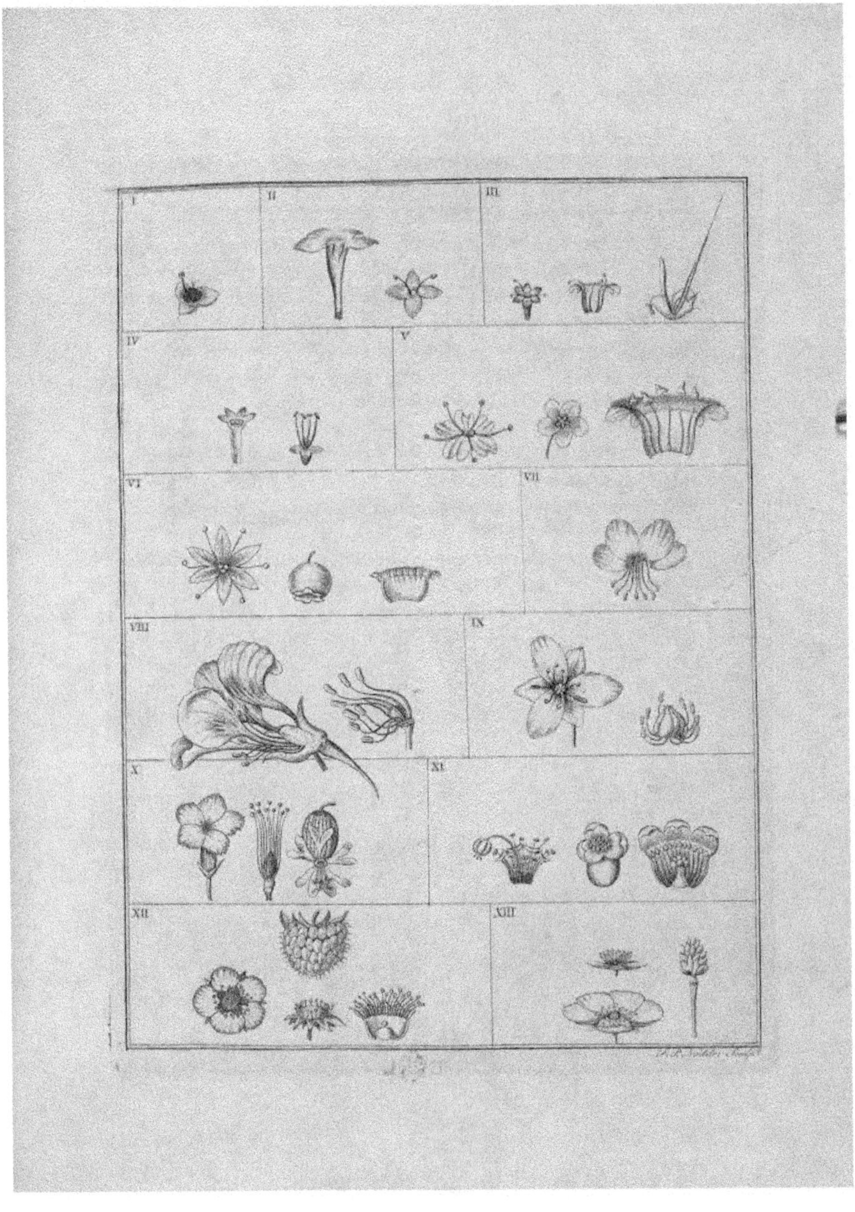

LP3

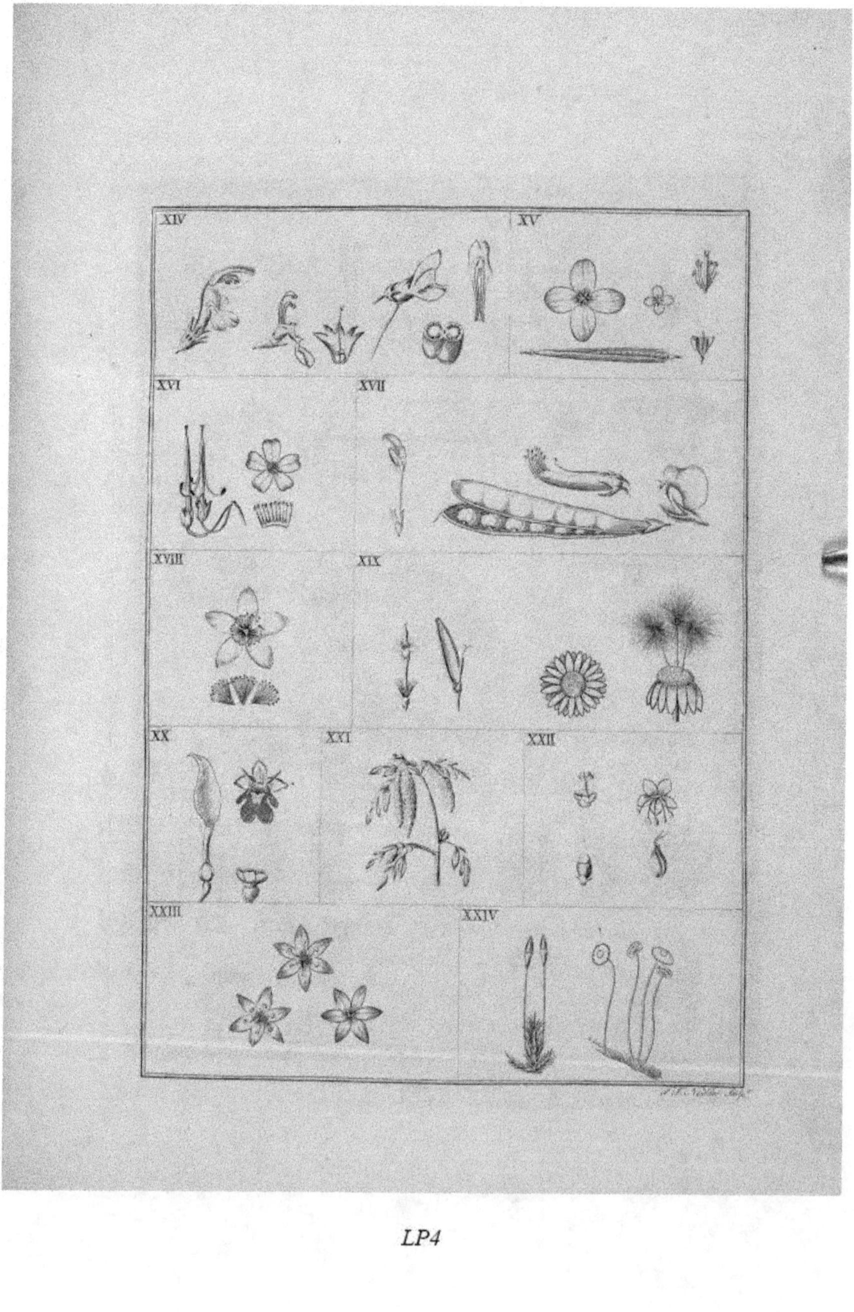

LP4

THE LOVES OF THE PLANTS.
CANTO I.

Descend, ye hovering Sylphs! aerial Quires,
And sweep with little hands your silver lyres;
With fairy foot-steps print your grassy rings,
Ye Gnomes! accordant to the tinkling strings;
While in soft notes I tune to oaten reed
Gay hopes, and amorous sorrows of the mead.—
From giant Oaks, that wave their branches dark,
To the dwarf Moss, that clings upon their bark,
What Beaux and Beauties croud the gaudy groves,
And woo and win their vegetable Loves.
How Snow-drops cold, and blue-eyed Harebels blend
Their tender tears, as o'er the stream they bend;
The love-sick Violet, and the Primrose pale
Bow their sweet heads, and whisper to the gale;
With secret sighs the Virgin Lily droops,
And jealous Cowslips hang their tawny cups.
How the young Rose in beauty's damask pride
Drinks the warm blushes of his bashful bride;
With honey'd lips enamour'd Woodbines meet,
Clasp with fond arms, and mix their kisses sweet.—

Stay thy soft-murmuring waters, gentle Rill;
Hush, whispering Winds, ye rustling Leaves, be still;
Rest, silver Butterflies, your quivering wings;
Alight, ye Beetles, from your airy rings;
Ye painted Moths, your gold-eyed plumage furl,

Vegetable Loves. l. 10. Linneus the celebrated Swedish naturalist, has demonstrated, that all flowers contain families of male or females, or both; and on their marriages has constructed his invaluable system of Botany.

Bow your wide horns, your spiral trunks uncurl;
Glitter, ye Glow-worms, on your mossy beds;
Descend, ye Spiders, on your lengthen'd threads;
Slide here, ye horned Snails, with varnish'd shells;
Ye Bee-nymphs, listen in your waxen cells!—

BOTANIC MUSE! who in this latter age
Led by your airy hand the Swedish sage,
Bad his keen eye your secret haunts explore
On dewy dell, high wood, and winding shore;
Say on each leaf how tiny Graces dwell;
How laugh the Pleasures in a blossom's bell;
How insect-Loves arise on cob-web wings,
Aim their light shafts, and point their little stings.

"First the tall CANNA lifts his curled brow
Erect to heaven, and plights his nuptial vow;
The virtuous pair, in milder regions born,
Dread the rude blast of Autumn's icy morn;
Round the chill fair he folds his crimson vest,
And clasps the timorous beauty to his breast.

Thy love, CALLITRICHE, *two* Virgins share,
Smit with thy starry eye and radiant hair; —
On the green margin sits the youth, and laves
His floating train of tresses in the waves;
Sees his fair features paint the streams that pass,
And bends for ever o'er the watery glass.

Two brother swains of COLLIN's gentle name,
The same their features, and their forms the same,

Canna. l. 39. Cane, or indian Reed. One male and one female inhabit each flower. It is brought from between the tropics to our hot-houses, and bears a beautiful crimson flower; the seeds are used as shot by the Indians, and are strung for prayer-beads in some catholic countries.

Callitriche. l. 45. Fine-Hair, Stargrass. One male and two females inhabit each flower. The upper leaves grow in form of a star, whence it is called Stellaria Aquatica by Ray and others; its stems and leaves float far on the water, and are often so matted together, as to bear a person walking on them. The male sometimes lives in a separate flower.

Collinsonia. l. 51. Two males one female. I have lately observed a very singular circumstance in this flower; the two males stand widely diverging from each other, and the female bends herself

LP5

With rival love for fair Collinia sigh,
Knit the dark brow, and roll the unsteady eye.
With sweet concern the pitying beauty mourns,
And sooths with smiles the jealous pair by turns.

Sweet blooms Genista in the myrtle shade,
And *ten* fond brothers woo the haughty maid.
Two knights before thy fragrant altar bend,
Adored Melissa! and *two* squires attend.
Meadia's soft chains *five* suppliant beaux confess,

---

into contact first with one of them, and after some time leaves this, and applies herself to the other. It is probable one of the anthers may be mature before the other? See note on Gloriosa, and Genista. The females in Nigella, devil in the bush, are very tall compared to the males; and bending over in a circle to them, give the flower some resemblance to a regal crown. The female of the epilobium angustifolium, rose bay willow herb, bends down amongst the males for several days, and becomes upright again, when impregnated.

Genista. l. 57. Dyer's broom. Ten males and one female inhabit this flower. The males are generally united at the bottom in two sets, whence Linneus has named the class "two brotherhoods." In the Genista, however, they are united in but one set. The flowers of this class are called papilionaceous, from their resemblance to a butterfly, as the pea-blossom. In the Spartium Scoparium, or common broom, I have lately observed a curious circumstance, the males or stamens are in two sets, one set rising a quarter of an inch above the other; the upper set does not arrive at their maturity so soon as the lower, and the stigma, or head of the female, is produced amongst the upper or immature set; but as soon as the pistil grows tall enough to burst open the keel-leaf, or hood of the flower, it bends itself round in an instant, like a french horn, and inserts its head, or stigma, amongst the lower or mature set of males. The pistil, or female, continues to grow in length; and in a few days the stigma arrives again amongst the upper set, by the time they become mature. This wonderful contrivance is readily seen by opening the keel-leaf of the flowers of broom before they burst spontaneously. See note on Collinsonia, Gloriosa, Draba.

Melissa. l. 60. Balm. In each flower there are four males and one female; two of the males stand higher than the other two; whence the name of the class "two powers." I have observed in the Ballota, and others of this class, that the two lower stamens, or males, become mature before the two higher. After they have shed their dust, they turn themselves away outwards; and the pistil, or female, continuing to grow a little taller, is applied to the upper stamens. See Gloriosa, and Genista.

All the plants of this class, which have naked seeds, are aromatic. The Marum, and Nepeta are particularly delightful to cats; no other brute animals seem pleased with any odours but those of their food or prey.

Meadia. l. 61. Dodecatheon, american Cowslip. Five males and one female. The males, or anthers, touch each other. The uncommon beauty of this flower occasioned Linneus to give it a name signifying the twelve heathen gods; and Dr. Mead to affix his own name to it. The pistil is much longer than the stamens, hence the flower-stalks have their elegant bend, that the stigma may hang downwards to receive the fecundating dust of the anthers. And the petals are so beautifully turned back to prevent the rain or dew drops from sliding down and washing off this dust prematurely; and at the same time exposing it to the light and air. As soon as the seeds are formed, it erects all the flower-stalks to prevent them from falling out; and thus loses the beauty of its figure. Is this a mechanical effect, or does it indicate a vegetable storgé to preserve its offspring? See note on Ilex, and Gloriosa.

And hand in hand the laughing belle address;
Alike to all, she bows with wanton air,
Rolls her dark eye, and waves her golden hair.

Woo'd with long care, CURCUMA cold and shy
Meets her fond husband with averted eye:
Four beardless youths the obdurate Beauty move
With soft attentions of Platonic love.

With vain desires the pensive ALCEA burns,
And, like sad ELOISA, loves and mourns.

In the Meadia, the Borago, Cyclamen, Solanum, and many others, the filaments are very short compared with the style. Hence it became necessary, 1st. to furnish the stamens with long anthers. 2d. To lengthen and bend the peduncle or flower-stalk, that the flower might hang downwards. 3d. To reflect the petals. 4th. To erect these peduncles when the germ was fecundated. We may reason upon this by observing, that all this aparatus might have been spared, if the filaments alone had grown longer; and that thence in these flowers that the filaments are the most unchangeable parts; and that thence their comparative length, in respect to the style, would afford a most permanent mark of their generic character.

Curcuma. l. 65. Turmeric. One male and one female inhabit this flower; but there are besides four imperfect males, or filaments without anthers upon them, called by Linneus eunuchs. The flax of our country has ten filaments, and but five of them are terminated with anthers; the Portugal flax has ten perfect males, or stamens; the Verbena of our country has four males; that of Sweden has but two; the genus Albuca, the Bignonia Catalpa, Gratiola, and hemlock-leaved Geranium have only half their filaments crowned with anthers. In like manner the florets, which form the rays of the flowers of the order frustraneous polygamy, of the class syngenesia, or confederate males, as the sun-flower, are furnished, with a style only, and no stigma; and are thence barren. There is also a style without a stigma in the whole order dioecia gynandria; the male flowers of which are thence barren. The Opulus is another plant, which contains some unprolific flowers. In like manner some tribes of insects have males, females, and neuters among them: as bees, wasps, ants.

There is a curious circumstance belonging to the class of insects which have two wings, or diptera, analagous to the rudiments of stamens above described; viz. two little knobs are found placed each on a stalk or peduncle, generally under a little arched scale; which appear to be rudiments of hinder wings; and are called by Linneus, halteres, or poisers, a term of his introduction. A. T. Bladh. Amæn. Acad. V. 7. Other animals have marks of having in a long process of time undergone changes in some parts of their bodies, which may have been effected to accommodate them to new ways of procuring their food. The existance of teats on the breasts of male animals, and which are generally replete with a thin kind of milk at their nativity, is a wonderful instance of this kind. Perhaps all the productions of nature are in their progress to greater perfection? an idea countenanced by the modern discoveries and deductions concerning the progressive formation of the solid parts of the terraqueous globe, and consonant to the dignity of the Creator of all things.

Alcea. l. 69. Flore pleno. Double hollyhock. The double flowers, so much admired by the florists, are termed by the botanist vegetable monsters; in some of these the petals are multiply'd three or four times, but without excluding the stamens, hence they produce some seeds, as Campanula and Stramoneum; but in others the petals become so numerous as totally to exclude the stamens,

The freckled IRIS owns a fiercer flame,
And *three* unjealous husbands wed the dame.
CUPRESSUS dark disdains his dusky bride,
One dome contains them, but *two* beds divide.
The proud OSYRIS flies his angry fair,
Two houses hold the fashionable pair.

With strange deformity PLANTAGO treads,
A Monster-birth! and lifts his hundred heads;
Yet with soft love a gentle belle he charms,
And clasps the beauty in his hundred arms.

---

or males; as Caltha, Peonia, and Alcea: these produce no seeds and are termed eunuchs. Philos. Botan. No. 150.

These vegetable monsters are formed in many ways. 1st. By the multiplication of the petals and the exclusion of the nectaries, as in larkspur. 2d. By the multiplication of the nectaries and exclusion of the petals; as in columbine. 3d. In some flowers growing in cymes, the wheel-shape flowers in the margin are multiplied to the exclusion of the bell-shape flowers in the centre; as in gelder rose. 4th. By the elongation of the florets in the centre. Instances of both these are found in daisy and feverfew; for other kinds of vegetable monsters, see Plantago.

The perianth is not changed in double flowers, hence the genus or family may be often discovered by the calyx, as in Hepatica, Ranunculus, Alcea. In those flowers, which have many petals, the lowest series of the petals remains unchanged in respect to number; hence the natural number of the petals is easily discovered. As in poppies, roses, and Nigella, or devil in a bush. Phil. Bot. p. 128.

Iris. l. 71. Flower de Luce. Three males, one female. Some of the species have a beautifully freckled flower; the large stigma or head of the female, covers the three males, counterfeiting a petal with its divisions.

Cupressus. l. 73. Cypress. One House. The males live in separate flowers, but on the same plant. The males of some of these plants, which are in separate flowers from the females, have an elastic membrane; which disperses their dust to a considerable distance, when the anthers burst open. This dust, on a fine day, may often be seen like a cloud hanging round the common nettle. The males and females of all the cone-bearing plants are in separate flowers, either on the same or different plants; they produce resins, and many of them are supposed to supply the most durable timber, what is called Venice-turpentine is obtained from the larch by wounding the bark about two feet from the ground, and catching it as it exsudes; Sandarach is procured from common juniper; and incense from a juniper with yellow fruit. The unperishable chests, which contain the Egyptian mummies, were of Cypress; and the Cedar, with which black-lead pencils are covered, is not liable to be eaten by worms. See Miln's Bot. Dict. art. coniferæ.

Osyris. l. 75. Two houses. The males and females are on different plants. There are many instances on record, where female plants have been impregnated at very great distance from their male; the dust discharged from the anthers, is very light, small, and copious, so that it may spread very wide in the atmosphere, and be carried to the distant pistils, without the supposition of any particular attraction; these plants resemble some insects, as the ants, and cochineal insect, of which the males have wings, but not the female.

Plantago. l. 77. Rosea. Rose-Plantain. In this vegetable monster the bractes, or divisions of the spike, become wonderfully enlarged; and are converted into leaves. The chaffy scales of the calyx in Xeranthemum, and in a species of Dianthus, and the glume in some alpine grasses, and the scales of the ament in the salix rosea, rose-willow, grow into leaves; and produce other kinds of monsters. The double flowers become monsters by the multiplication of their petals or nectaries. See note on Alcea.

So hapless DESDEMONA, fair and young,
Won by OTHELLO's captivating tongue,
Sigh'd o'er each strange and pitious tale, distress'd,
And sunk enamour'd on his sooty breast.

Two gentle shepherds and their sister-wives
With thee, ANTHOXA! lead ambrosial lives;
Where the wide heath in purple pride extends,
And scatter'd furze its golden lustre blends,
Closed in a green recess, unenvy'd lot!
The blue smoak rises from their turf-built cot;
Bosom'd in fragrance blush their infant train,
Eye the warm fun, or drink the silver rain.

The fair OSMUNDA seeks the silent dell,
The ivy canopy, the dripping cell;
There hid in shades *clandestine* rites approves,
Till the green progeny betrays her loves.

With charms despotic fair CHONDRILLA reigns
O'er the soft hearts of *five* fraternal swains;

Anthoxanthum. 1.86. Vernal grass. Two males, two females. The other grasses have three males and two females. The flowers of this grass give the fragrant scent to hay. I am informed it is frequently viviparous, that is, that it bears sometimes roots or bulbs instead of seeds, which after a time drop off and strike root into the ground. This circumstance is said to obtain in many of the alpine grasses, whose seeds are perpetually devoured by small birds. The Festuca Dumetorum, fescue grass of the bushes, produces bulbs from the sheaths of its straw. The Allium Magicum, or magical onion, produces onions on its head, instead of seeds. The Polygonum Viviparum, viviparous bistort, rises about a foot high, with a beautiful spike of flowers, which are succeeded by buds or bulbs, which fall off and take root. There is a bush frequently seen on birch-trees, like a bird's nest, which seems to be a similar attempt of nature, to produce another tree; which falling off might take root in spongy ground.

There is an instance of this double mode of production in the animal kingdom, which is equally extraordinary: the same species of Aphis is viviparous in summer, and oviparous in autumn. A. T. Bladh. Amæn. Acad. V. 7.

Osmunda. l. 93. This plant grows on moist rocks; the parts of its flower or its seeds are scarce discernible; whence Linneus has given the name of clandestine marriage to this class. The younger plants are of a beautiful vivid green.

Chondrilla. l. 97. Of the class Confederate Males. The numerous florets, which constitute the disk of the flowers of this class, contain in each five males surrounding one female, which are connected at top, whence the name of the class. An Italian writer in a disscourse on the irritability of flowers asserts, that if the top of the floret be touched, all the filaments which support the cylindrical anther will contract themselves, and that by thus raising or depressing the anther the whole of the prolific dust is collected on the stigma. He adds, that if one filament be touched after

Gloriosa Superba. Engraved by Fredk. P. Nodder

LP6

If sighs the changeful nymph, alike they mourn;
And, if she smiles, with rival raptures burn.
So, tuned in unison, Eolian Lyre!
Sounds in sweet symphony thy kindred wire;
Now, gently swept by Zephyr's vernal wings,
Sink in soft cadences thy love-sick strings;
And now with mingling chords, and voices higher,
Peal the full anthems of the aerial choir.

Five sister-nymphs to join Diana's train
With thee, fair LYCHNIS! vow,—but vow in vain;
Beneath one roof resides the virgin band,
Flies the fond swain, and scorns his offer'd hand;
But when soft hours on breezy pinions move,
And smiling May attunes her lute to love,
Each wanton beauty, trick'd in all her grace,
Shakes the bright dew-drops from her blushing face;
In gay undress displays her rival charms,
And calls her wondering lovers to her arms.

When thve young Hours amid her tangled hair
Wove the fresh rose-bud, and the lily fair,
Proud GLORIOSA led *three* chosen swains,
The blushing captives of her virgin chains.—

---

it is separated from the floret, that it will contract like the muscular fibres of animal bodies, his experiements were tried on the Centauréa Calcitrapoides, and on artichokes, and globe-thistles. Discourse on irritability of plants. Dodsley.

Lychnis. l. 108. Ten males and five females. The flowers, which contain the five females, and those which contain the ten males, are found on different plants; and often at a great distance from each other. Five of the ten males arrive at their maturity some days before the other five, as may be seen by opening the corol before it naturally expands itself. When the females arrive at their maturity, they rise above the petals, as if looking abroad for their distant husbands, the scarlet ones contribute much to the beauty of our meadows in May and June.

Gloriosa. l. 119. Superba. Six males one female. The petals of this beautiful flower with three of the stamens, which are first mature, stand up in apparent disorder; and the pistil bends at nearly a right angle to insert its stigma amongst them. In a few days, as these decline, the other three stamens bend over, and approach the pistil. In the Fritillaria Persica, the six stamens are of equal lengths, and the anthers lie at a distance from the pistil, and three alternate ones approach first; and, when these decline, the other three approach; in the Lithrum Salicaria, (which has twelve males and one female) a beautiful red flower, which grows on the banks of rivers, six of the males arrive at maturity, and surround the female sometime before the other six; when these decline, the other six rise up, and supply their places. Several other flowers have in similar manner two sets of

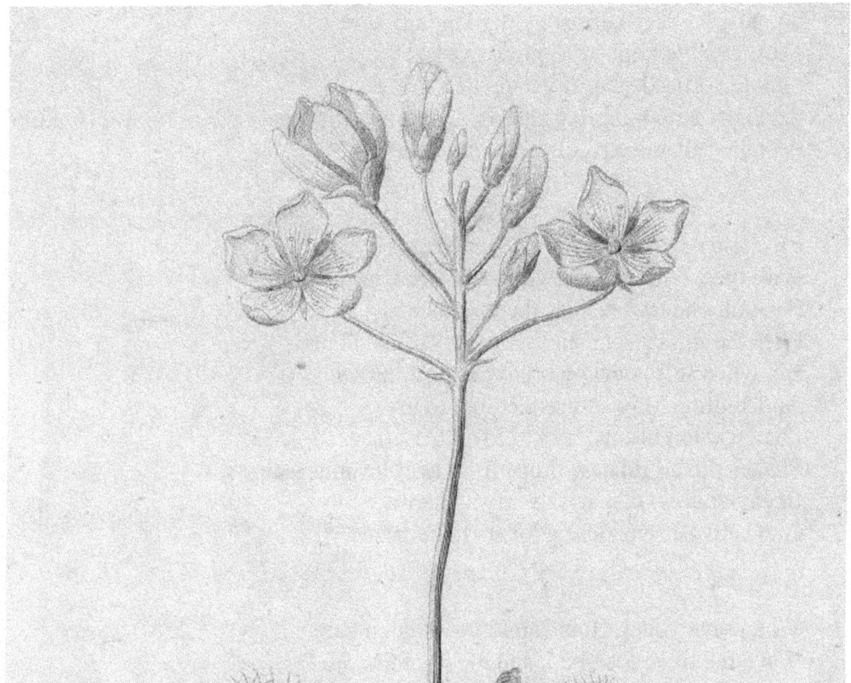

LP7

——When Time's rude hand a bark of wrinkles spread
Round her weak limbs, and silver'd o'er her head,
Three other youths her riper years engage,
The flatter'd victims of her wily age.
So NINON pruned her wither'd charms, and won
With harlot-smiles her gay unconscious son;—
Clasp'd in his arms she own'd a mother's name,
Shook her grey locks, and tittering mock'd his flame;
With mad despair he plunged the guilty dart,
And life and love gush'd mingled from his heart!

The fell SILENE and her sisters fair,
Skill'd in destruction, spread the viscous snare.
The harlot-band *ten* lofty bravoes screen,
And frowning guard the magic nets, unseen.—
Haste, glittering nations, tenants of the air,
Oh, steer from hence your viewless course afar!
If with soft words, sweet blushes, nods, and smiles,
The *three* dread Sirens lure you to their toils,
Limed with their art in vain you point your stings,
In vain the efforts of your whirring wings!—
Go, seek your gilded mates and infant hives,
Nor taste the honey purchas'd with your lives!

Four of the giant brood with ILEX stand,
Each grasps a thousand arrows in his hand;

---

stamina of different ages, as Adoxa, Lychnis, Saxifraga. See Genista. Perhaps a difference in the time of their maturity obtains in all those flowers, which have numerous stamens.

Silene. l. 131. Catchfly. Three females and ten males inhabit each flower; the viscous material, which surrounds the stalks under the flowers of this plant, and of the Cucubulus Otites, is a curious contrivance to prevent various insects from plundering the honey, or devouring the seed. In the Dionæa Muscipula there is a still more wonderful contrivance to prevent the depredations of insects: The leaves are armed with long teeth, like the antennæ of insects, and lie spread upon the ground round the stem; and are so irritable, that when an insect creeps upon them, they fold up, and crush or pierce it to death. The last professor Linneus in his supplementum plantarum, gives the following account of the Arum Muscivorum. The flower has the smell of carrion; by which the flies are invited to lay their eggs in the chamber of the flower, but in vain endeavour to escape, being prevented by the hairs pointing inwards; and thus perish in the flower, whence its name of fly-eater. P. 411, in the Dypsacus is another contrivance for this purpose, a bason of water is placed round each joint of the stem. In the Drosera is another kind of fly trap. See Dypsacus and Drosera; the flowers of Siléne and Cucúbalus are closed all day, but are open and give an agreeable odour in the night. See Cerea.

Ilex. l. 143. Holly. Four males, four females. Many plants, like many animals, are furnished with arms for their protection; these are either aculei, prickles, as in rose and barberry, which are formed from the outer bark of the plant; or spinæ, thorns, as in hawthorn, which are an elongation

A thousand steely points on every scale
Form the bright terrors of his bristly mail.—
So arm'd, immortal Moore uncharmed the spell,
And slew the wily dragon of the well.—
Sudden with rage their *injur'd* bosoms burn,
Retort the insult, or the wound return;
Unwrong'd, as gentle as the breeze, that sweeps
The unbending harvests or undimpled deeps,
They guard, the Kings of Needwood's wide domains,
Their sister-wives and fair infantine trains;
Lead the lone pilgrim through the trackless glade,
Or guide in leafy wilds the wandring maid.

Gigantic Nymph! the fair KLEINHOVIA reigns
The grace and terror of her wide domains;
O'er her warm cheek the blush of beauty swims,
And nerves herculean bend her sinewy limbs;
With frolic eye she views the affrighted throng,
And shakes the meadows, as she towers along,
With playful violence displays her charms,

---

of the wood, and hence more difficult to be torn off than the former; or stimuli, stings, as in the nettles, which are armed with a venomous fluid for the annoyance of naked animals. The shrubs and trees, which have prickles or thorns, are grateful food to many animals, as gooseberry, and gorse; and would be quickly devoured, if not thus armed; the stings seem protection against some kinds of insects, as well as the naked mouths of quadrupeds. Many plants lose their thorns by cultivation, as wild animals lose their ferocity; and some of them their horns. A curious circumstance attends the large Hollies in Needwood-forest, they are armed with thorny leaves about eight feet high, and have smooth leaves above; as if they were conscious, that horses and cattle could not reach their upper branches. See notes on Meadia, and on Mancinella. The numerous clumps of hollies in Needwood forest serve as land-marks to direct the travellers across it in various directions; and as a shelter to the deer and cattle in winter; and in scarce seasons supply them with much food. For when the upper branches, which are without prickles, are cut down, the deer crop the leaves and peel off the bark. The bird-lime made from the bark of Hollies seems to be a very similar material to the elastic gum, or indian rubber, as it is called.

Kleinhovia. l. 157. In this class the males in each flower are supported by the female. The name of the class may be translated "Viragoes." Or "Feminine Males."

The largest tree perhaps in the world is of the same natural order as Kleinhovia, it is the Adansonia, or Ethiopian Sour-gourd, or African Calabash tree. Mr. Adanson says the diameter of the trunk frequently exceeds 25 feet, and the horizontal branches are from 45 to 55 feet long, and so large that each branch is equal to the largest trees of Europe. The breadth of the top is from 120 to 150 feet. And one of the roots bared only in part by the washing away of the earth by the river, near which it grew, measured 110 feet long; and yet these stupendous trees never exceed 70 feet in height. Voyage to Senegal.

And bears her puny lovers in her arms.
So fair THALESTRIS bound her jutting breast
In rigid mail, and shook her plumy crest;—
Poised her long lance amid the walks of war,
And Beauty thunder'd from Bellona's car;
Greece arm'd in vain, her captive heroes wove
The chains of conquest with the wreaths of love.

When o'er the cultured lawns and dreary wastes
Retiring Autumn flings her howling blasts,
Bends in tumultuous waves the struggling woods,
And showers their leafy honours on the floods,
In withering heaps collects the flowery spoil,
And each chill insect sinks beneath the soil;
Quick flies fair TULIPA the loud alarms,
And folds her infant closer in her arms;
In some lone cave, secure pavilion, lies,
And waits the courtship of serener skies.—
So to his mossy couch the Dormouse springs,
And Sleep protects him with his eider wings.—
Bright out of earth, amid the driving storm,
Ascends fair COLCHICA! thy roseate form;

Tulipa. l. 177. Tulip. What is in common language called a bulbous root, is by Linneus termed the Hybernacle, or Winter-lodge of the young plant. As these bulbs in every respect resemble buds, except in their being produced under ground, and include the leaves and flower in miniature, which are to be expanded in the ensusing spring. By cautiously cutting through the concentric coats of a tulip-root, longitudinally from the top to the base, and taking them off successively, the whole flower of next year's tulip is beautifully seen by the naked eye, with its petals, pistil, and stamens; the flowers exist in other bulbs, in the same manner, as in Hyacinths, but the individual flowers of these being less, they are not so easily dissected, or so conspicuous to the naked eye.

In the seeds of the Nymphæa Nelumbo, the leaves of the plant are seen so distinctly, that Mr. Ferber found out by them to what plant the seeds belonged. Amæn. Acad. V. vi. No. 120. He says that Mariotte first observed the future flower and foliage in the bulb of a Tulip; and adds that it is pleasant to see in the buds of the Hepatica, and Pedicularis hirsuta, yet lying in the earth; and in the gems of Daphne Mezereon; and at the base of Osmunda Lunaria, a perfect plant of the future year compleat in all its parts. ibid.

Colchicum autumnale. l.184. Autumnal Meadow-saffron. Six males, three females. The germ is buried within the root, which thus seems to constitute a part of the flower. Families of Plants. p. 242. These singular flowers appear in the autumn without any leaves, whence in some countries they are called Naked Ladies: in the March following the green leaves spring up, and in April the seed-vessel rises from the ground; the seeds ripen in May, contrary to the usual habits of vegetables, which flower in the spring, and ripen their seeds in the autumn. Miller's Dict. The juice of the root of this plant is so acrid as to produce violent effects on the human constitution, which also prevents it from being eaten by subterranean insects, and thus guards the seed-vessel during the winter. The defoliation of deciduous trees is announced by the flowering of the

Warms the cold bosom of the hoary year,
And lights with Beauty's blaze the dusky sphere.
Three blushing Maids the intrepid Nymph attend,
And *six* gay Youths, enamour'd train! defend.
So shines with silver guards the Georgian star,
And drives o'er Night's blue arch his glittering car.

Great HELIANTHUS guides o'er twilight plains
In gay solemnity his Dervice-trains;
Marshall'd in *fives* each gaudy band proceeds,
Each gaudy band a plumed Lady leads;
With zealous step he climbs the upland lawn,
And bows in homage to the rising dawn;
Imbibes with eagle-eye the golden ray,
And watches, as it moves, the orb of day.

Queen of the dewy vale, fair DROSERA treads
Her moss-wove banks, and rushy-fringed beds;

Colchicum; of these the ash is the last that puts forth its leaves, and the first that looses them. Phil. Bot. p. 275.

The Hamamelis, Witch Hazle, is another plant which flowers in autumn; when the leaves fall off, the flowers come out in clusters from the joints of the branches, and in Virginia ripen their seed in ensuing spring; but in this country their seeds seldom ripen. Lin. Spec. Plant. Miller's Dict.

Helianthus. l. 191. Sun flower. The numerous florets, which constitute the disk of this flower, contain in each five males surrounding one female, the five stamens have their anthers connected at top, whence the name of the class "confederate males;" see note on Chondrilla. The sun flower follows the course of the sun by nutation, not by twisting its stem. (Hales veg. stat.) Other plants, when they are confined in a room, turn the shining surface of their leaves, and bend their whole branches to the light. See Mimosa.

A plumed lady leads. l. 194. The seeds of many plants of this class are furnished with a plume, by which admirable mechanism they are disseminated by the winds far from their parent stem, and look like a shuttlecock, as they fly. Other seeds are disseminated by animals, of these some attach themselves to their hair or feathers by a gluten, as mistleto; others by hooks, as clevers, burdock, houndstongue; and others are swallowed whole for the sake of the fruit, and voided uninjured, as the hawthorn, juniper, and some grasses. Other seeds again disperse themselves by means of an elastic seed-vessel, as Oats, Geranium, and Impatiens; and the seeds of aquatic plants, and of those, which grow on the banks of rivers, are carried many miles by the currents, into which they fall. See Impatiens. Zostera. Cassia.

Drosera. l. 199. Sun-dew. Five males, five females. The leaves of this marsh-plant are purple, and have a fringe very unlike other vegetable productions. And, which is curious, at the point of every thread of this erect fringe stands a pellucid drop of mucilage, resembling a ducal coronet. This mucus is a secretion from certain glands, and like the viscous material round the flower-stalks of Silene, (catchfly) prevents small insects from infesting the leaves. As the ear-wax in animals seems to be in part designed to prevent fleas and other insects from getting into their ears. See Silene. Mr. Wheatly an eminent surgeon in Cateaton-street, London, observed these leaves

Redundant folds of glossy silk surround
Her slender waist, and trail upon the ground;
Five sister-nymphs collect with graceful ease,
Or spread the floating purple to the breeze;
And *five* fair youths with duteous love comply
With each soft mandate of her moving eye.
As with sweet grace her snowy neck she bows,
A zone of diamonds trembles round her brows;
Bright shines the silver halo, as she turns;
And, as she steps, the living lustre burns.

Fair LONICERA treads the dewy lawn,
And decks with brighter blush the vermil dawn;
Winds round the shadowy rocks, and pansied vales,
And scents with sweeter breath the summer-gales;
With artless grace and native ease she charms,
And bears the Horn of Plenty in her arms.

---

to bend upwards when an insect settled on them, like the leaves of the muscipula veneris, and pointing all their globules of mucus to the centre, that they compleatly intangled and destroyed it.

Lonicera. l. 211. Caprifolium. Honeysuckle. Five males, one female. Nature has in many flower used a wonderful apparatus to guard the nectary, or honey-gland, from insects. In the honeysuckle the petal terminates in a long tube like a cornucopiæ, or horn of plenty; and the honey is produced at the bottom of it. In Aconitum, monks-hood, the nectaries stand upright like two horns covered with a hood, which abounds with such acrid matter that no insects penetrate it. In Helleborus, hellebore, the many nectaries are placed in a circle, like little pitchers, and add much to the beauty of the flower. In the Columbine, Aquilegia, the nectary is imagined to be like the neck and body of a bird, and the two petals standing upon each side to represent wings; whence its name of columbine, as if resembling a nest of young pigeons fluttering whilst their parent feeds them. The importance of the nectary in the economy of vegetation is explained at large in the notes on part the first.

Many insects are provided with a long and pliant probiscis for the purpose of acquiring this grateful food, as a variety of bees, moths, and butterflies: but the Sphinx Convolvuli, or unicorn moth, is furnished with the most remarkable probiscis in this climate. It carries it rolled up in concentric circles under its chin, and occasionally extends it to above three inches in length. This trunk consists of joints and muscles, and seems to have more versatile movements than the trunk of the elephant; and near its termination is split into two capillary tubes. The excellence of this contrivance for robbing the flowers of their honey, keeps this beautiful insect fat and bulky; though it flies only in the evening, when the flowers have closed their petals, and are thence more difficult of access; at the same time the brilliant colours of the moth contribute to its safety, by making it mistaken by the late sleeping birds for the flower it rests on.

Besides these there is a curious contrivance attending the Ophrys, commonly called the Bee-orchis, and the Fly-orchis, with some kinds of the Delphinium, called Bee-larkspurs, to preserve their honey; in these the nectary and petals resemble in form and colour the insects, which plunder them: and thus it may be supposed, they often escape these hourly robbers, by having the appearance of being pre-occupied. See note on Rubia, and Conferva polymorpha.

Five rival Swains their tender cares unfold,
And watch with eye askance the treasured gold.

Where rears huge Tenerif his azure crest,
Aspiring Draba builds her eagle nest;
Pleased round the fair *four* rival Lords ascend
The shaggy steeps, *two* menial youths attend.
High in the setting ray the beauty stands,
And her tall shadow waves on distant lands.

Stay, bright inhabitant of air, alight,
Ambitious Visca, from thy eagle-flight!—
——Scorning the sordid soil, aloft she springs,
Shakes her white plume, and claps her golden wings;
High o'er the fields of boundless ether roves,
And seeks amid the clouds her soaring loves!

Stretch'd on her mossy couch, in trackless deeps,
Queen of the coral groves, Zostera sleeps;

Draba. 1.220. Alpina. Alpine Whitlow-grass. One female and six males. Four of these males stand above the other two; whence the name of the class "four powers." I have observed in several plants of this class, that the two lower males arise, in a few days after the opening of the flower, to the same height as the other four, not being mature as soon as the higher ones. See note on Gloriosa. All the plants of this class possess similar virtues; they are termed acrid and antiscorbutic in their raw state. As mustard, watercress; when cultivated and boiled, they become a mild wholesome food, as cabbage, turnep.

Viscum. l. 226. Misletoe. Two houses. This plant never grows upon the ground; the foliage is yellow, and the berries milk-white; the berries are so viscous, as to serve for bird-lime; and when they fall, adhere to the branches of the tree, on which the plant grows, and strike root into its bark; or are carried to distant trees by birds. The Tillandsia, or wild pine, grows on other trees, like the Misletoe, but takes little or no nourishment from them, having large buckets in its leaves to collect and retain the rain water. See note on Dypsacus. The mosses, which grow on the bark of trees, take much nourishment from them, hence it is observed that trees, which are annually cleared from moss by a brush, grow nearly twice as fast. (Phil. transact.) In the cyder countries the peasants brush their apple-trees annually.

Zostera. l.232. Grass-wrack. Class, Feminine Males. Order, Many Males. It grows at the bottom of the sea, and rising to the surface, when in flower, covers many leagues; and is driven at length to the shore. During its time of floating on the sea, numberless animals live on the under surface of it; and being specifically lighter than the sea water, or being repelled by it, have legs placed as it were on their backs for the purpose of walking under it. As the Scyllœa. See Barbut's genera vermium. It seems necessary that the marriages of plants should be celebrated in the open air, either because the powder of the anther, or the mucilage on the stigma, or the reservoir

The silvery sea-weed matted round her bed,
And distant surges murmuring o'er her head.—
High in the flood her azure dome ascends,
The crystal arch on crystal columns bends;
Roof'd with translucent shell the turrets blaze,
And far in ocean dart their colour'd rays;
O'er the white floor successive shadows move,
As rise and break the ruffled waves above.—
Around the nymph her mermaid-trains repair,
And weave with orient pearl her radiant hair;
With rapid fins she cleaves the watery way,
Shoots like a silver meteor up to day;
Sounds a loud conch, convokes a scaly band,
Her sea-born lovers, and ascends the strand.

E'en round the pole the flames of Love aspire,
And icy bosoms feel the *secret* fire!—
Cradled in snow and fan'd by arctic air
Shines, gentle BAROMETZ! thy golden hair;

---

of honey might receive injury from the water. Mr. Needham observed, that in the ripe dust of every flower, examined by the microscope, some vesicles are perceived, from which a fluid had escaped; and that those, which still retain it, explode if they be wetted, like an eolopile suddenly exposed to a strong heat. These observations have been verified by Spallanzani, and others. Hence rainy seasons make a scarcity of grain, or hinder its fecundity, by bursting the pollen before it arrives at the moist stigma of the flower. Spallanzani's dissertations, v.2. p. 321. Thus the flowers of the male Vallisneria are produced under water, and when ripe detach themselves from the plant, and rising to the surface are wafted by the air, to the female flowers. See Vallisneria.

Barometz. l. 250. Polypodium Barometz. Tartarian Lamb. Clandestine Marriage. This species of Fern is a native of China, with a decumbent root, thick, and every where cover'd with the most soft and dense wool, intensely yellow. Lin. Spec. Plant.

This curious stem is sometimes push'd out of the ground in its horizontal situation by some of the inferior branches of the root, so as to give it some resemblance to a Lamb, standing on four legs; and has been said to destroy all other plants in its vicinity. Sir Hans Sloane describes it under the name of Tartarian Lamb, and has given a print of it. Philos. Trans. abridg'd, V. 2. p. 646. but thinks some art had been used to give it an animal appearance. Dr. Hunter in his edition of the Terra of Evelyn, has given a more curious print of it, much resembling a sheep. The down is used in India externally for stopping hemorrhages, and is called golden moss.

The thick downy clothing of some vegetables seems designed to protect them from the injuires of cold, like the wool of animals. Those bodies which are bad conductors of electricity, are also bad conductors of heat, as glass, wax, air. Hence either of the two former of these may be melted by the flame of a blow-pipe very near the fingers which hold it, without burning them; and the last, by being confined on the surface of animal bodies, in the interstices of their fur or wool, prevents the escape of their natural warmth; to which should be added, that the hairs themselves are imperfect conductors. The fat or oil of whales, and other northern animals, seems designed for the same purpose of preventing the too sudden escape of the heat of the body in cold climates.

Rooted in earth each cloven hoof descends,
And round and round her flexile neck she bends;
Crops the grey coral moss, and hoary thyme,
Or laps with rosy tongue the melting rime;
Eyes with mute tenderness her distant dam,
Or seems to bleat, a *Vegetable Lamb*.
—So, warm and buoyant in his oily mail,
Gambols on seas of ice the unwieldy Whale;
Wide-waving fins round floating islands urge
His bulk gigantic through the troubled surge;
With hideous yawn the flying shoals He seeks,
Or clasps with fringe of horn his massy cheeks;
Lifts o'er the tossing wave his nostrils bare,
And spouts pellucid columns into air;
The silvery arches catch the setting beams,
And transient rainbows tremble o'er the streams.

Weak with nice sense, the chaste Mimosa stands,
From each rude touch withdraws her timid hands;
Oft as light clouds o'erpass the Summer-glade,
Alarm'd she trembles at the moving shade;

Snow protects vegetables, which are cover'd by it from cold, both because it is a bad conductor of heat itself, and contains much air in its pores. If a piece of camphor be immersed in a snow-ball except one extremity of it, on setting fire to this, as the snow melts the water becomes absorbed into the surrounding snow by capillary attraction; on this account when living animals are burry'd in snow, they are not moistened by it; but the cavity enlarges as the snow dissolves, affording them both a dry and warm habitation.

Mimosa. l. 267. The sensitive plant. Of the class Polygamy, one house. Naturalists have not explained the immediate cause of the collapsing of the sensitive plant; the leaves meet and close in the night during the sleep of the plant, or when exposed to much cold in the day time, in the same manner as when they are affected by external violence, folding their upper surfaces together, and in part over each other like scales or tiles; so as to expose as little of the upper surface as may be to the air; but do not indeed collapse quite so far, since I have found, when touched in the night during their sleep, they fall still further; especially when touched on the foot-stalks between the stems and the leaflets, which seems to be their most sensitive or irritable part. Now as their situation after being exposed to external violence resembles their sleep, but with a greater degree of collapse, may it not be owing to a numbness or paralysis consequent to too violent irritation, like the faintings of animals from pain or fatigue? I kept a sensitive plant in a dark room till some hours after day break, its leaves and leaf-stalks were collapsed as in its most profound sleep, and on exposing it to the light, above twenty minutes passed before the plant was thoroughly awake and had quite expanded itself. During the night the upper or smoother surfaces of the leaves are appressed together, this would seem to shew that the office of this surface of the leaf was to expose the fluids of the plant to the light as well as to the air. See note on Helianthus. Many flowers close up their petals during the night. See note on vegetable respiration in part 1st.

And feels, alive through all her tender form,
The whisper'd murmurs of the gathering storm;
Shuts her sweet eye-lids to approaching night;
And hails with freshen'd charms the rising light.
Veil'd, with gay decency and modest pride,
Slow to the mosque she moves, an eastern bride;
There her soft vows unceasing love record,
Queen of the bright seraglio of her Lord.—
So sinks or rises with the changeful hour
The liquid silver in its glassy tower.
So turns the needle to the pole it loves,
With fine librations quivering, as it moves.

All wan and shivering in the leafless glade
The sad ANEMONE reclin'd her head;
Grief on her cheeks had paled the roseate hue,
And her sweet eye-lids drop'd with pearly dew.
—"See, from bright regions, borne on odorous gales
The Swallow, herald of the summer, sails;

Anemone. l. 284. Many males, many females. Pliny says, this flower never opens its petals, but when the wind blows, whence its name; it has properly no calix, but two or three sets of petals, three in each set, which are folded over the stamens and pistil in a singular and beautiful manner, and differs also from ranunculus in not having a melliferous pore on the claw of each petal.

The Swallow. l. 288.There is a wonderful conformity between the vegetation of some plants, and the arrival of certain birds of passage. Linneus observes that the wood anemone blows in Sweden on the arrival of the swallow; and the marsh mary-gold, Caltha, when the cuckoo sings. Nearly the same coincidence was observed in England by Stillingfleet. The word Coccux in greek signifies both a young fig and a cuckoo, which is supposed to have arisen from the coincidence of their appearance in Greece. Perhaps a similar coincidence of appearance in some parts of Asia gave occasion to the story of the loves of the rose and nightingale so much celebrated by the eastern poets. See Dianthus. The times however of the appearance of vegetables in the spring seem occasionally to be influenced by their acquired habits, as well as by their sensibility to heat: for the roots of potatoes, onions, &c. will germinate with much less heat in the spring, than in the autumn; as is easily observable, where these roots are stored for use; and hence malt is best made in the spring. 2d. The grains and roots brought from more southern latitudes germinate here sooner than those which are brought from more northern ones, owning to their acquired habits. Fordyce on agriculture. 3d. It was observed by one of the scholars of Linneus, that the apple-trees sent from hence to new England blossomed for a few years too early for that climate, and bore no fruit; but afterwards learnt to accommodate themselves to their new situation. (Kalm's travels.) 4th. The parts of animals become more sensible to heat after having been previously exposed to cold, as our hands glow on coming into the house after having held snow in them; this seems to happen to vegetables, for vines in grape-houses, which have been exposed to the winter's cold, will become forwarder and more vigorous than those, which have been kept during the winter in the house. (Kenedy on Gardening.) This accounts for the very rapid vegetation in the northern latitudes, after the solution of the snows.

The increase of the irritability of plants in respect to heat, after having been previously exposed to cold, is further illustrated by an experiment of Dr. Walker's. He cut apertures into a birch-tree

Breathe, gentle AIR! from cherub-lips impart
Thy balmy influence to my anguish'd heart;
Thou, whose soft voice calls forth the tender blooms,
Whose pencil paints them, and whole breath perfumes;
O chase the Fiend of Frost, whose leaden mace
In death-like slumbers seals my hapless race;
Melt his hard heart, release his iron hand,
And give my ivory petals to expand.
So may each bud, that decks the brow of spring,
Shed all its incense on thy wafting wing!"—
To her fond prayer propitious Zephyr yields,
Sweeps on his sliding shell through azure fields,
O'er her fair mansion waves his whispering wand,
And gives her ivory petals to expand:
Gives with new life her filial train to rise,
And hail with kindling smiles the genial skies.
So shines the Nymph in beauty's blushing pride,
When Zephyr wafts her deep calash aside;
Tears with rude kiss her bosoms gauzy veil,
And flings the fluttering kerchief to the gale.
So bright, the folding canopy undrawn,
Glides the gilt Landau o'er the velvet lawn,
Of beaux and belles displays the glittering throng;
And soft airs fan them, as they roll along.

Where frowning Snowden bends his dizzy brow
O'er Conway, listening to the surge below;
Retiring LICHEN climbs the topmost stone,
And 'mid the airy ocean dwells alone.—
Bright shine the stars unnumber'd *o'er her head*,
And the cold moon-beam gilds her flinty bed;
While round the rifted rocks hoarse whirlwinds breathe,

---

at different heights; and on the 26th of March some of these apertures bled, or oozed with the sap-juice, when the thermometer was at 39; which same apertures did not bleed on the 13th of March when the thermometer was at 44. The reason of this I apprehend was, because on the night of the 25th the thermometer was as low as 34, whereas on the night of the 12th it was at 41; tho' the ingenious author ascribes it to another cause. Trans. of Royal Soc. of Edinburgh. V. I. p. 19.

Lichen. l. 315. Calcareum. Liver-wort. Clandestine Marriage. This plant is the first that vegetates on naked rocks, covering them with a kind of tapestry, and draws its nourishment perhaps chiefly from the air; after it perishes, earth enough is left for other mosses to root themselves; and after some ages a soil is produced sufficient for the growth of more succulent and large vegetables. In this manner perhaps the whole earth has been gradually covered with vegetation, after it was raised out of the primeval ocean by subterraneous fires.

And dark with thunder sail the clouds *beneath*.—
The steepy path her plighted swain pursues,
And tracks her light step o'er the imprinted dews;
Delighted Hymen gives his torch to blaze,
Winds round the craggs, and lights the mazy ways;
Sheds o'er their secret vows his influence chaste,
And decks with roses the admiring waste.

High in the front of heaven when Sirius glares,
And o'er Britannia shakes his fiery hairs;
When no soft shower descends, no dew distills,
Her wave-worn channels dry, and mute her rills;
When droops the sickening herb, the blossom fades,
And parch'd earth gapes beneath the withering glades.
——With languid step fair DYPSACA retreats;
"Fall gentle dews!" the fainting nymph repeats;
Seeks the low dell, and in the sultry shade
Invokes in vain the Naiads to her aid.—
Four sylvan youths in crystal goblets bear
The untasted treasure to the grateful fair;
Pleased from their hands with modest grace she sips,
And the cool wave reflects her coral lips.

With nice selection modest RUBIA blends
Her vermil dyes, and o'er the cauldron bends;

Dypsacus. l. 333. Teasel. One female, and four males. There is a cup around every joint of the stem of this plant, which contains from a spoonful to half a pint of water; and serves both for the nutriment of the plant in dry seasons, and to prevent insects from creeping up to devour its seed. See Silene. The Tillandsia, or wild pine, of the West Indies has every leaf terminated near the stalk with a hollow bucket, which contains from half a pint to a quart of water. Dampier's voyage to Campeachy. Dr. Sloane mentions one kind of aloe furnished with leaves, which like the wild pine, and Banana, hold water; and thence afford necessary refreshment to travelers in hot countries. Nepenthes has a bucket for the same purpose at the end of every leaf. Burm. Zeyl. 42. 17.

Rubia. l. 341. Madder. Four males and one female. This plant is cultivated in very large quantities for dying red. If mixed with the food of young pigs or chickens, it colours their bones red. If they are fed alternate fortnights with a mixture of madder, and with their usual food alone, their bones will consist of concentric circles of white and red. Belchier. Phil. trans. 1736. Animals fed with madder for the purpose of these experiments were found upon dissection to have thinner gall. Comment. de rebus. Lipsiæ, Vol. . This circumstance is worth further attention. The colouring materials of vegetables, like those which serve the purpose of tanning, varnishing, and the various medical purposes, do not seem essential to the life of the plant; but seem given it as a defence against the depredations of insects or other animals, to whom these materials are nauseous or deleterious. To insects and many smaller animals their colours contribute to conceal them from the larger ones which pray upon them. Caterpillars which feed on leaves, are

Warm 'mid the rising steam the beauty glows,
As blushes in a mist the dewy rose.
With chemic art *four* favour'd youths aloof
Stain the white fleece, or stretch the tinted woof;
O'er Age's cheek the warmth of youth diffuse,
Or deck the pale-eyed nymph in roseate hues.
So when MEDEA to exulting Greece
From plundered COLCHIS bore the golden fleece;
On the loud shore a magic pile she rais'd,
The cauldron bubbled, and the faggots blaz'd;—
Pleased on the boiling wave old ÆSON swims,
And feels new vigour stretch his swelling limbs;
Through his thrill'd nerves forgotten ardors dart,
And warmer eddies circle round his heart;
With softer fires his kindling eye-balls glow,
And locks luxuriant wanton round his brow.

As dash the waves on India's breezy strand,
Her flush'd cheek press'd upon her lily hand,

generally green, and earth-worms the colour of the earth which they inhabit; Butterflies, which frequent flowers, are coloured like them; small birds which frequent hedges, have greenish backs like the leaves, and light coloured bellies, like the sky, and are hence less visible to the hawk, who passes under them or over them. Those birds which are much amongst flowers, as the gold-finch, (Fringilla carduelis) are furnished with vivid colours. The lark, partridge, hare, are the colour of the dry vegetables or earth, on which they rest. And frogs vary their colour with the mud of the streams which they frequent, and those which live on trees are green. Fish, which are generally suspended in water; and swallows, which are generally suspended in air, have their backs the colour of the distant ground, and their bellies of the sky. In the colder climates many of these become white during the existence of the snows. Hence there is apparent design in the colours of animals, whilst those of vegetables seem consequent to the other properties of the materials which possess them.

Pleased on the boiling wave. l. 353. The story of Aeson's becoming young, from the medicated bath of Medea, seems to have been intended to teach the efficacy of warm bathing in retarding the progress of old age. The words *relaxation and bracing*, which are generally thought expressive of the effects of warm and cold bathing, are mechanical terms, properly applied to drums or strings; but are only metaphors, when applied to the effects of cold or warm bathing on animal bodies. The immediate cause of old age seems to reside in the inirritability of the finer vessels or parts of our system; hence these cease to act, and collapse or become horny or bony. The warm bath is peculiarly adapted to prevent these circumstances by its increasing our irritability, and by moistening and softening the skin, and the extremities of the finer vessels, which terminate in it. To those, who are past the meridian of life, and have dry skins, and begin to be emaciated, the warm bath, for half an hour twice a week, I believe to be eminently serviceable in retarding the advances of age.

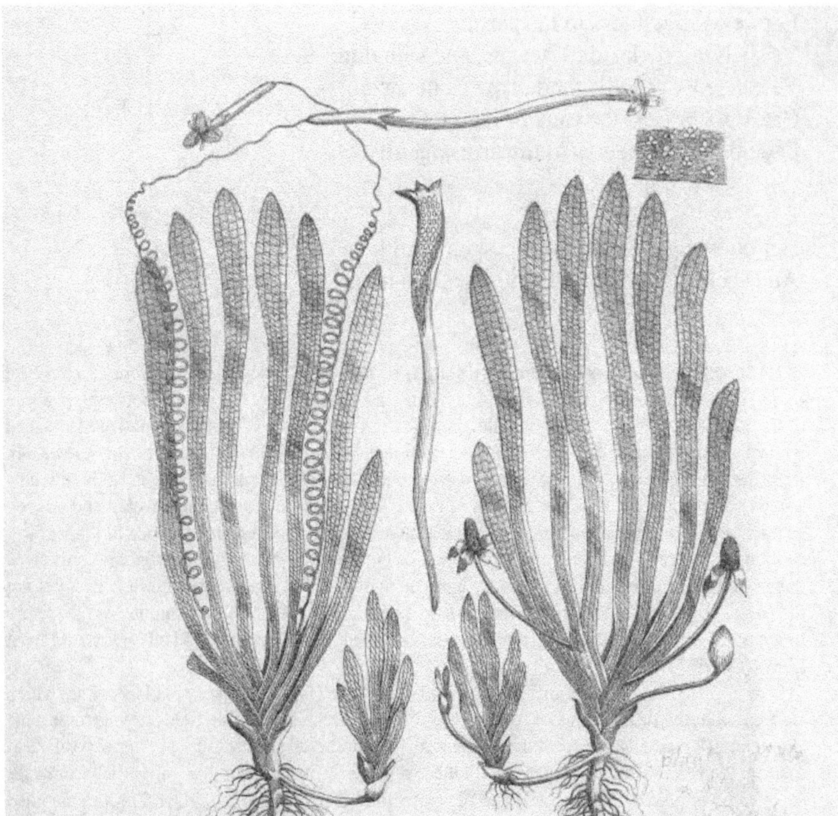

LPS

Vallisner sits, upturns her tearful eyes,
Calls her lost lover, and upbraids the skies;
For him she breathes the silent sigh, forlorn,
Each setting day; for him each rising morn.—
"Bright orbs, that light yon high ethereal plain,
Or bathe your radiant tresses in the main;
Pale moon, that silver'st o'er night's sable brow;—
For ye were witness to his parting vow!—
Ye shelving rocks, dark waves, and sounding shore,—
Ye echoed sweet the tender words he swore!—
Can stars or seas the sails of love retain?
O guide my wanderer to my arms again!"

Her buoyant skiff the intrepid Ulva guides,
And seeks her Lord amid the trackless tides;

Vallisneria. l. 361. This extraordinary plant is of the class Two Houses. It is found in the East Indias, in Norway and various parts of Italy. Lin. Spec. Plant. They have their roots at the bottom of the Rhone, the flowers of the female plant float on the surface of the water, and are furnished with an elastic spiral stalk, which extends or contracts as the water rises and falls; this rise or fall, from the rapid descent of the river, and the mountain torrents which flow into it, often amounts to many feet in a few hours. The flowers of the male plant are produced under water, and as soon as their farina, or dust, is mature; they detach themselves from the plant, and rise to the surface, continue to flourish, and are wafted by the air, or borne by the currents to the female flowers. In this resembling those tribes of insects, where the males at certain seasons acquire wings, but not the females, as ants. These male flowers are in such numbers, though very minute, as frequently to cover the surface of the river to considerable extent. See Families of Plants translated from Linneus, p. 677.

Ulva. l. 373. Clandestine marriage. This kind of sea-weed is bouyed up by bladders of air, which are formed in the duplicatures of its leaves; and forms immense floating fields of vegetation; the young ones, branching out from the larger ones, and borne on similar little air-vessels. It is also found in the warm baths of Patavia; where the leaves are formed into curious cells or labyrinths for the purpose of floating on the water. See ulva labyrinthi-formis Lin. Spec. Plant. The air contained in these cells was found by Dr. Priestly to be sometimes purer than common air, and sometimes less pure; the air-bladders of fish seem to be similar organs, and serve to render them buoyant in the water. In some of these, as in the Cod and Haddock, a red membrane consisting of a great number of leaves or duplicatures is found within the air-bag, which probably secretes this air from the blood of the animal. (Monro. Physiol. of Fish. p. 28.) to determine whether this air, when first separated from the blood of the animal or plant, be dephlogisticated air, is worthy inquiry. The bladder-sena, (Colutea,) and bladder-nut (Staphylæa) have their seed-vessels distended with air; the Ketmia has the upper joint of the stem immediately under the receptacle of the flower, much distended with air; these seem to be analogous to the air-vessels distended with air; these seem to be analogous to the air-vessel at the broad end of the egg, and may probably become less pure as the seed ripens: some, which I tried, had the purity of the surrounding atmosphere. The air at the broad end of the egg is probably an organ serving the purpose of respiration to the young chick, some of whose vessels are spread upon it like a placenta, or permeate it. Many are of opinion that even the placenta of the human fetus, and cotyledons of quadrupeds are respiratory organs rather than nutritious ones.

Her *secret* vows the Cyprian Queen approves,
And hovering halcyons guard her infant-loves;
Each in his floating cradle round they throng,
And dimpling Ocean bears the fleet along.—
Thus o'er the waves, which gently bend and swell,
Fair GALATEA steers her silver shell;
Her playful Dolphins stretch the silken rein,
Hear her sweet voice, and glide along the main.
As round the wild meandering coast she moves
By gushing rills, rude cliffs, and nodding groves;
Each by her pine the Wood-nymphs wave their locks,
And wondering Naiads peep amid the rocks;
Pleased trains of Mermaids rise from coral cells,
Admiring Tritons sound their twisted shells;
Charm'd o'er the car pursuing Cupids sweep,
Their snow-white pinions twinkling in the deep;
And, as the lustre of her eye she turns,
Soft sighs the Gale, and amorous Ocean burns.

On DOVE's green brink the fair TREMELLA stood,
And view'd her playful image in the flood;

The air in the hollow stems of grasses, and of some umbelliferous plants, bears analogy to the air in the quills, and in some of the bones of birds; supplying the place of the pith, which shrivels up after it has performed its office of protruding the young stem or feather. Some of these cavities of the bones are said to communicate with the lungs in birds. Phil. Trans.

The air-bladders of fish are nicely adapted to their intended purpose, for though they render them buoyant near the surface, without the labour of using their fins; yet when they rest at greater depths, they are no inconvenience, as the increased pressure of the water condenses the air, which they contain, into less space. Thus if a cork or a bladder of air was immersed a very great depth in the ocean, it would be so much compressed, as to become specifically as heavy as the water, and would remain there. It is probable the unfortunate Mr. Day, who was drowned in a diving ship of his own construction, miscarried from not attending to this circumstance: it is probable the quantity of air he took down with him, if he descended much lower than he expected, was condensed into so small a space as not to render the ship buoyant, when he endeavoured to ascend.

Tremella. l. 393. Clandestine marriage. I have frequently observed fungusses of this Genus on old rails and on the ground to become a transparent jelly, after they had been frozen in autumnal mornings; which is a curious property, and distinguishes them from some other vegetable mucilage; for I have observed that the paste, made by boiling wheat-flour in water, ceases to be adhesive after having been frozen. I suspected that the Tremella Nostoc, or star-jelly, also had been thus produced; but have since been well informed, that the Tremella Nostoc is a mucilage voided by Herons after they have eaten frogs; hence it has the appearance of having been pressed through a hole; and limbs of frogs are said sometimes to be found amongst it; it is always seen upon plains or by the sides of water, places which Herons generally frequent.

To each rude rock, lone dell, and echoing grove
Sung the sweet sorrows of her *secret* love.
"Oh, stay!—return!"—along the sounding shore
Cry'd the sad Naiads,—she return'd no more!—
Now girt with clouds the sullen Evening frown'd,
And withering Eurus swept along the ground;
The misty moon withdrew her horned light,
And sunk with Hesper in the skirt of night;
No dim electric streams, (the northern dawn,)
With meek effulgence quiver'd o'er the lawn;
Nor star benignant shot one transient ray
To guide or light the wanderer on her way.
Round the dark craggs the murmuring whirlwinds blow,
Woods groan above, and waters roar below;
As o'er the steeps with pausing foot she moves,
The pitying Dryads shriek amid their groves;
She flys,—she stops,—she pants—she looks behind,
And hears a demon howl in every wind.
—As the bleak blast unfurls her fluttering vest,
Cold beats the snow upon her shuddering breast;
Through her numb'd limbs the chill sensations dart,
And the keen ice-bolt trembles at her heart.
"I sink, I fall! oh, help me, help!" she cries,
Her stiffening tongue the unfinish'd sound denies;
Tear after tear adown her cheek succeeds,
And pearls of ice bestrew the glistering meads;
Congealing snows her lingering feet surround,
Arrest her flight, and root her to the ground;
With suppliant arms she pours the silent prayer,
Her suppliant arms hang crystal in the air;

Some of the Funguses are so acrid, that a drop of their juice blisters the tongue; others intoxicate those who eat them. The Ostiahs in Siberia use them for the latter purpose; one Fungus of the species, Agaricus muscarum, eaten raw; or the decoction of three of them, produces intoxication for 12 to 16 hours. History of Russia. V. 1. Nichols. 1780. As all acrid plants become less so, if exposed to a boiling heat, it is probable the common mushroom may sometimes disagree from being not sufficiently stewed. The Ostiacks blister their skin by a fungus found on Birch-trees; and use the Agaricus officin. for Soap. ib.

There was a dispute whether the funguses should be classed in the animal or vegetable department. Their animal taste in cookery, and their animal smell when burnt, together with their tendency to putrefaction, insomuch that the Phallus impudicus has gained the name of stink-horn, and lastly their growing and continuing healthy without light, as the fungus vinosus in dark cellars, and the esculent mushrooms on beds covered thick with straw, would seem to shew, that they approach towards the animals, or make a kind of isthmus connecting the two mighty kingdoms of animal and of vegetable nature.

Pellucid films her shivering neck o'erspread,
Seal her mute lips, and silver o'er her head,
Veil her pale bosom, glaze her lifted hands,
And shrined in ice the beauteous statue stands.
—Dove's azure nymphs on each revolving year
For fair Tremella shed the tender tear;
With rush-wove crowns in sad procession move,
And sound the sorrowing shell to hapless love.

Here paused the Muse,—across the darken'd pole
Sail the dim clouds, the echoing thunders roll;
The trembling Wood-nymphs, as the tempest lowers,
Lead the gay Goddess to their inmost bowers;
Hang the mute lyre the laurel shade beneath,
And round her temples bind the myrtle wreath.
—Now the light swallow with her airy brood
Skims the green meadow, and the dimpled flood;
Loud shrieks the lone thrush from his leafless thorn,
Th' alarmed beetle sounds his bugle horn;
Each pendant spider winds with fingers fine
His ravel'd clue, and climbs along the line;
Gay Gnomes in glittering circles stand aloof
Beneath a spreading mushroom's fretted roof;
Swift bees returning seek their waxen cells,
And Sylphs cling quivering in the lily's bells.
Through the still air descend the genial showers,
And pearly rain-drops deck the laughing flowers.

INTERLUDE.

Bookseller. Your verses, Mr. Botanist, consist of *pure description*, I hope there 1
is *sense* in the notes.

Poet. I am only a flower-painter, or occasionally attempt a landskip; and leave 2
the human figure with the portraits of history to abler artists.

B. It is well to know what subjects are within the limits of your pencil; many 3
have failed of success from the want of this self-knowledge. But pray tell me, what is the essential difference between Poetry and Prose? is it solely the melody or measure of the language?

P. I think not solely; for some prose has its melody, and even measure. And 4
good verses, well spoken in a language unknown to the hearer, are not easily to be distinguished from good prose.

B. Is it the sublimity, beauty, or novelty of the sentiments? 5

P. Not so, for sublime sentiments are often better expressed in prose. Thus when 6
Warwick in one of the plays of Shakespear is left wounded on the field after the loss of the battle, and his friend says to him, "Oh, could you but fly!" what can be more sublime than his answer, "Why then, I would not fly." No measure of verse I imagine could add dignity to this sentiment. And it would be easy to select examples of the beautiful or new from prose writers, which I suppose no measure of verse could improve.

B. In what then consists the essential difference between Poetry and Prose? 7

P. Next to the measure of the language, the principal distinction appears to me to 8
consist in this; that Poetry admits of very few words expressive of perfectly abstracted ideas, whereas Prose abounds with them. And as our ideas derived from visible objects are more distinct than those derived from the objects of our other senses, the words expressive of these ideas belonging to vision make up the principal part of poetic language. That is the Poet writes principally to the eye, the Prose-writer uses more abstracted terms. Mr. Pope has written a bad verse in the Windsor Forest.

"And Kennet swift for silver Eels *renown'd*." 9

The word renown'd does not present the idea of a visible object to the mind, and 10
is thence prosaic. But change this line thus,

"And Kennet swift, where silver Graylings *play*." 11

and it becomes poetry, because the scenery is then brought before the eye. 12

B. This may be done in prose. 13

P. And when it is done in a single word, it animates the prose; so it is more agree- 14
able to read in Mr. Gibbon's history, "Germany was at this time *over-shadowed* with extensive forests." Than Germany was at this time *full* of extensive forests. But where this mode of expression occurs too frequently, the prose approaches to poetry: and in graver works, where we expect to be instructed rather than amused, it becomes tedious and impertinent. Some parts of Mr. Burk's eloquent orations become intricate and enervated by superfluity of poetic ornament; which quantity of ornament would have been agreeable in a poem, where much ornament is expected.

B. Is then the office of Poetry only to amuse? 15

P. The Muses are young ladies, we expect to see them dressed; though not like 16
some modern beauties with so much gauze and feather, that "the Lady herself is the least part of her." There are however didactic pieces of poetry, which are much admired, as the Georgics of Virgil, Mason's English Garden, Hayley's Epistles; nevertheless Science is best delivered in Prose, as its mode of reasoning is from stricter analogies than metaphors or similies.

B. Do not Personifications and Allegories distinguish poetry? 17

P. These are other arts of bringing objects before the eye; or of expressing 18
sentiments in the language of vision; and are indeed better suited to the pen than the pencil.

B. That is strange, when you have just said they are used to bring their objects 19
before the eye.

P. In poetry the personification or allegoric figure is generally indistinct, and 20
therefore does not strike us so forceably as to make us attend to its improbability; but in painting, the figures being all much more distinct, their improbability becomes apparent, and seizes our attention to it. Thus the person of Concealment is very indistinct, and therefore does not compel us to attend to its improbability, in the following beautiful lines of Shakespear.

"——She never told her love; 21
But let Concealment, like a worm i' th' bud, 22
Feed on her damask cheek."— 23

But in these lines below the person of Reason obtrudes itself into our company, 24
and becomes disagreeable by its distinctness, and consequent improbability.

"To Reason I flew, and intreated her aid, 25
Who paused on my case, and each circumstance weigh'd; 26
Then gravely reply'd in return to my prayer, 27
That Hebe was fairest of all that were fair. 28
That's a truth reply'd I, I've no need to be taught, 29

I came to you, Reason, to find out a fault. 30
If that's all, says Reason, return as you came, 31
To find fault with Hebe would forfeit my name." 32

Allegoric figures are on this account in general less manageable in painting and 33
in statuary than in poetry: and can seldom be introduced in the two former arts in company with natural figures, as is evident from the ridiculous effect of many of the paintings of Rubens in the Luxemburgh gallery; and for this reason, because their improbability becomes more striking, when there are the figures of real persons by their side to compare them with.

Mrs. Angelica Kauffman, well apprised of this circumstance, has introduced 34
no mortal figures amongst her Cupids and her Graces. And the great Roubiliac, in his unrivalled monument of Time and Fame struggling for the trophy of General Flemming, has only hung up a medallion of the head of the hero of the piece. There are however some allegoric figures, which we have so often heard described or seen delineated, that we almost forget, that they do not exist in common life; and hence view them without astonishment; as the figures of the heathen mythology, of angels, devils, death and time; and almost believe them to be realities, even when they are mixed with representations of the natural forms of man. Whence I conclude, that a certain degree of probability is necessary to prevent us from revolting with distaste from unnatural images; unless we are otherwise so much interested in the contemplation of them as not to perceive their improbability.

B. Is this reasoning about degrees of probability just?—When Sir Joshua Reyn- 35
olds, who is unequaled both in the theory and practice of his art, and who is a great master of the pen as well as the pencil, has asserted in a discourse delivered to the Royal Academy, Decem. 11, 1786, that "the higher styles of painting, like the higher kinds of the Drama, do not aim at any thing like deception; or have any expectation, that the spectators should think the events there represented are really passing before them." And he then accuses Mr. Fielding of bad judgment, when he attempts to compliment Mr. Garrick in one of his novels, by introducing an ignorant man, mistaking the representation of a scene in Hamlet for a reality; and thinks, because he was an ignorant man, he was less liable to make such a mistake.

P. It is a metaphysical question, and requires more attention than Sir Joshua has 36
bestowed upon it.—You will allow, that we are perfectly deceived in our dreams; and that even in our waking reveries, we are often so much absorbed in the contemplation of what passes in our imaginations, that for a while we do not attend to the lapse of time or to our own locality; and thus suffer a similar kind of deception as in our dreams. That is, we believe things present before our eyes, which are not so.

There are two circumstances, which contribute to this compleat deception in our 37
dreams. First, because in sleep the organs of sense are closed or inert, and hence the trains of ideas associated in our imaginations are never interrupted or dissevered by the irritations of external objects, and can not therefore be contrasted with our sensations. On this account, tho' we are affected with a variety of passions in

our dreams, as anger, love, joy; yet we never experience surprize.—For surprize is only produced, when any external irritations suddenly obtrude themselves, and dissever our passing trains of ideas.

Secondly, because in sleep there is a total suspension of our voluntary power, 38
both over the muscles of our bodies, and the ideas of our minds; for we neither walk about, nor reason in compleat sleep. Hence, as the trains of ideas are passing in our imaginations in dreams, we cannot compare them with our previous knowledge of things, as we do in our waking hours; for this is a voluntary exertion, and thus we cannot perceive their incongruity.

Thus we are deprived in sleep of the only two means by which we can distin- 39
guish the trains of ideas passing in our imaginations, from those excited by our sensations; and are led by their vivacity to believe them to belong to the latter. For the vivacity of these trains of ideas, passing in the imagination, is greatly increased by the causes above-mentioned; that is, by their not being disturbed or dissevered either by the appulses of external bodies, as in surprize; or by our voluntary exertions in comparing them with our previous knowledge of things, as in reasoning upon them.

B. Now to apply. 40

P. When by the art of the Painter or Poet a train of ideas is suggested to our 41
imaginations, which interests us so much by the pain or pleasure it affords, that we cease to attend to the irritations of common external objects, and cease also to use any voluntary efforts to compare these interesting trains of ideas with our previous knowledge of things, a compleat reverie is produced: during which time however short, if it be but for a moment, the objects themselves appear to exist before us. This I think has been called by an ingenious critic "the ideal presence," of such objects. (Elements of Criticism by Lord Kaimes.) And in respect to the compliment intended by Mr. Fielding to Mr. Garrick, it would seem that an ignorant Rustic at the play of Hamlet, who has some previous belief in the appearance of Ghosts, would sooner be liable to fall into reverie, and continue in it longer, than one who possessed more knowledge of the real nature of things, and had a greater facility of exercising his reason.

B. It must require great art in the Painter or Poet to produce this kind of 42
deception.

P. The matter must be interesting from its sublimity, beauty, or novelty; this is 43
the scientific part; and the art consists in bringing these distinctly before the eye, so as produce (as above mentioned) the ideal presence of the object, in which the great Shakespear particularly excells.

B. Then it is not of any consequence, whether the representations correspond 44
with nature.

P. Not if they so much interest the reader or spectator as to induce the reverie 45
above described. Nature may be seen in the market place, or at the card-table; but we expect something more than this in the play-house or picture-room. The further the artist recedes from nature, the greater novelty he is likely to produce; if he rises above nature, he produces the sublime; and beauty is probably a selection and new

combination of her most agreeable parts. Yourself will be sensible of the truth of this doctrine by recollecting over in your mind the works of three of our celebrated artists. Sir Joshua Reynolds has introduced sublimity even into his portraits; we admire the representation of persons, whose reality we should have passed by unnoticed. Mrs. Angelica Kauffman attracts our eyes with beauty, which I suppose no where exists; certainly few Grecian faces are seen in this country. And the daring pencil of Fusseli transports us beyond the boundaries of nature, and ravishes us with the charm of the most interesting novelty. And Shakespear, who excels in all these together, so far captivates the spectator, as to make him unmindful of every kind of violation of Time, Place, or Existence. As at the first appearance of the Ghost of Hamlet, "his ear must be dull as the fat weed, which roots itself on Lethe's brink," who can attend to the improbability of the exhibition. So in many scenes of the Tempest we perpetually believe the action passing before our eyes, and relapse with somewhat of distaste into common life at the intervals of the representation.

B. I suppose a poet of less ability would find such great machinery difficult and 46
cumbersome to manage.

P. Just so, we should be shocked at the apparent improbabilities. As in the 47
gardens of a Scicilian nobleman, described in Mr. Brydone's travels, there are said to be six hundred statues of imaginary monsters; which so disgust the spectators, that the state had once a serious design of destroying them; and yet the very improbable monsters in Ovid's Metamorphoses have entertained the world for many centuries.

B. The monsters in your Botanic Garden, I hope are of the latter kind? 48

P. The candid reader must determine. 49

THE LOVES OF THE PLANTS.
CANTO II

AGAIN the Goddess strikes the golden lyre,
And tunes to wilder notes the warbling wire;
With soft suspended step Attention moves,
And Silence hovers o'er the listening groves;
Orb within orb the charmed audience throng,
And the green vault reverberates the song.

"Breathe soft, ye Gales!" the fair CARLINA cries,
"Bear on broad wings your Votress to the skies.
How sweetly mutable yon orient hues,
As Morn's red hand her opening roses strews;
How bright, when Iris blending many a ray
Binds in embroider'd wreath the brow of Day;
Soft, when the pendant Moon with lustres pale
O'er heaven's blue arch unfurls her milky veil;
While from the north long threads of silver light
Dart on swift shuttles o'er the tissued night!

Carlina. l. 7. Carline Thistle. Of the class Confederate Males. The seeds of this, and of many other plants of the same class, are furnished with a plume, by which admirable mechanism they perform long aerial journeys, crossing lakes and deserts, and are thus disseminated far from the original plant, and have much the appearance of a Shuttlecock, as they fly. The wings are of different construction, some being like a divergent tuft of hairs, others are branched like feathers, some are elevated from the crown of the seed by a slender foot-stalk, which gives them a very elegant appearance, others sit immediately on the crown of the seed.

Nature has many other curious vegetable contrivances for the dispersion of seeds, see note on Helianthus. But perhaps none of them has more the appearance of design than the admirable apparatus of Tillandsia for this purpose. This plant grows on the branches of trees, like the misleto, and never on the ground; the seeds are furnished with many long threads on their crowns. Which, as they are driven forwards by the winds, wrap round the arms of trees, and thus hold them fast till they vegetate. This is very analogous to the migration of Spiders on the gossamer, who are said to attach themselves to the end of a long thread, and rise thus to the tops of trees or buildings, as the accidental breezes carry them.

Breathe soft, ye Zephyrs! hear my fervent sighs,
Bear on broad wings your Votress to the skies!"—
—Plume over plume in long divergent lines
On Whale-bone ribs the fair Mechanic joins;
Inlays with eider down the silken strings,
And weaves in wide expanse Dædalean wings;
Round her bold sons the waving pennons binds,
And walks with angel-step upon the winds.—
So on the shoreless air the intrepid Gaul
Launch'd the vast concave of his buoyant ball.—
Journeying on high, the silken castle glides
Bright as a meteor through the azure tides;
O'er towns and towers and temples wins its way,
Or mounts sublime, and gilds the vault of day.
Silent with upturn'd eyes unbreathing crouds
Pursue the floating wonder to the clouds;
And flush'd with transport or benumb'd with fear
Watch, as it rises, the diminish'd sphere.
—Now less and less!—and now a speck is seen!—
And now the fleeting rack obtrudes between!—
With bended knees, raised arms, and suppliant brow
To every shrine with mingled cries they vow.—
"Save Him, ye Saints! who o'er the good preside;
Bear Him, ye Winds! ye Stars benignant! guide."
—The calm Philosopher in ether sails,
Views broader stars, and breathes in purer gales!
Sees, like a map, in many a waving line
Round Earth's blue plains her lucid waters shine;
Sees at his feet the forky lightenings glow,
And hears innocuous thunders roar below.
—Rise, great MONGULFIER! urge thy venturous flight
High o'er the Moon's pale ice-reflected light;
High o'er the pearly Star, whose beamy horn
Hangs in the east, gay harbinger of morn;
Leave the red eye of Mars on rapid wing,
Jove's silver guards, and Saturn's dusky ring;
Leave the fair beams, which issuing from afar
Play with new lustres round the Georgian star;
Shun with strong oars the Sun's attractive throne,
The burning zodiac, and the milky zone;
Where headlong Comets with increasing force
Through other systems bend their blazing course.—
For thee Cassiope her chair withdraws,

For thee the Bear retracts his shaggy paws,
High o'er the North thy golden orb shall roll,
And blaze eternal round the wondering pole.
So Argo, rising from the southern main,
Lights with new stars the blue ethereal plain;
With favoring beams the mariner protects,
And the bold course, which first it steer'd, directs.

Inventress of the Woof, fair LINA flings
The flying shuttle through the dancing strings;
Inlays the broider'd weft with flowery dyes,
Quick beat the reeds, the pedals fall and rise;
Slow from the beam the lengths of warp unwind,
And dance and nod the massy weights behind.—
Taught by her labours, from the fertile soil
Immortal ISIS clothed the banks of Nile;
And fair ARACHNE with her rival loom
Found undeserved a melancholy doom.—
Five Sister-nymphs with dewy fingers twine
The beamy flax, and stretch the fibre-line;
Quick eddying threads from rapid spindles reel,
Or whirl with beating foot the dizzy wheel.
—Charm'd round the busy Fair *five* shepherds press,
Praise the nice texture of their snowy dress,
Admire the Artists, and the art approve,
And tell with honey'd words the tale of love.

So now, where Derwent guides his dusky floods
Through vaulted mountains, and a night of woods,
The Nymph, GOSSYPIA, treads the velvet sod,

For thee the Bear. l. 60. Tibi jam brachia contrahit ardens Scorpius. Virg. Georg. l. I. 34.

Linum. l. 67. Flax. Five males and five females. It was first found on the banks of the Nile. The Linum Lusitanicum or portigal flax, has ten males, see the note on Curcuma. Isis was said to invent spinning and weaving: mankind before that time were clothed with the skins of animals. The fable of Arachne was to compliment this new art of spinning and weaving, supposed to surpass in fineness the web of the Spider.

Gossypia. l. 87. Gossypium. The cotton plant. On the river Derwent near Matlock in Derbyshire, SIR RICHARD ARKWRIGHT has erected his curious and magnificent machinery for spinning cotton; which had been in vain attempted by many ingenious artists before him. The cotton-wool is first picked from the pods and seeds by women. It is then carded by *cylindrical cards*, which move against each other, with different velocities. It is taken from these by an *iron-band* or comb, which has a motion similar to that of scratching, and takes the wool off the cards longitudinally in respect

And warms with rosy smiles the watery God;
His ponderous oars to slender spindles turns,
And pours o'er massy wheels his foamy urns;
With playful charms her hoary lover wins,
And wields his trident,—while the Monarch spins.
—First with nice eye emerging Naiads cull
From leathery pods the vegetable wool;
With wiry teeth *revolving cords* release
The tangled knots, and smooth the ravell'd fleece;
Next moves the *iron-band* with fingers fine,
Combs the wide card, and forms the eternal line;
Slow, with soft lips, the *whirling Can* acquires
The tender skeins, and wraps in rising spires;
With quicken'd pace successive rollers move,
And these retain, and those extend the *rove*;
Then fly the spoles, the rapid axles glow;—
And slowly circumvolves the labouring wheel below.

---

to the fibres or staple, producing a continued line loosely cohering called the *Rove* or *Roving*. This Rove yet very loosely twisted, is then received or drawn into a *whirling canister,* and is rolled by the centrifugal force in spiral lines within it; being yet too tender for the spindle. It is then passed between *two pairs of rollers*; the second pair moving faster than the first elongate the thread with greater equality, than can be done by the hand; and is then twisted on spoles or bobbins.

The great fertility of the Cotton-plant in these fine flexile threads, whilst those from Flax, Hemp, and Nettles, or from the bark of the Mulberry-tree, require a previous putrefaction of the parenchymatous substance, and much mechanical labour, and afterwards bleaching, renders this plant of great importance to the world. And since Sir Richard Arkwright's ingenious machine has not only greatly abreviated and simplefied the labour and art of carding and spinning the Cotton-wool; but performs both these circumstances *better* than can be done by hand; it is probable, that the clothing of this small seed will become the principal clothing of mankind. Though animal wool and silk may be preferable in colder climates, as they are more imperfect conductors of heat, and are thence a warmer clothing.

PAPYRA, throned upon the banks of Nile,
Spread her smooth leaf, and waved her silver style.
—The storied pyramid, the laurel'd bust,
The trophy'd arch had crumbled into dust;
The sacred symbol, and the epic song,
(Unknown the character, forgot the tongue,)
With each unconquer'd chief, or sainted maid,
Sunk undistinguish'd in Oblivion's shade.
Sad o'er the scatter'd ruins Genius sigh'd,
And infant Arts but learn'd to lisp and died.
Till to astonish'd realms PAPYRA taught
To paint in the mystic colours Sound and Thought.
With Wisdom's voice to print the page sublime,
And mark in adamant the steps of Time.
—*Three* favour'd youths with fond officious care
Learn the strange process, and assist the fair;
The first from Alpha to Omega joins
The letter'd tribes along the level lines;
Weighs with nice ear the vowel, liquid, surd,
And breaks in syllables the volant word.
Then forms *the next* upon the marshal'd plain
In deepening ranks his dexterous cypher-train;
And counts, as wheel the decimating bands,
The dews of Ægypt, or Arabia's sands.
And then *the third* on four concordant lines
Prints the lone crotchet, and the quaver joins;

Cyperus. Papyrus. l. 105. Three males, one female. The leaf of this plant was first used for paper, whence the word paper; and leaf, or folium, for a fold of a book. Afterwards the bark of a species of mulberry was used; whence liber signifies a book, and the bark of a tree. Before the invention of letters, mankind may be said to have been perpetually in their infancy, as the arts of one age or country generally died with their inventors. Whence arose the policy, which still continues in Indostan, of obliging the son to practice the profession of his father. After the discovery of letters, the facts of Astronomy and Chemistry became recorded in written language, though the antient hieroglyphic characters for the planets and metals continue in use to this day. The antiquity of the invention of music, of astronomical observations, and the manufacture of Gold and Iron are recorded in Scripture.

About twenty letters, ten ciphers, and seven crotchets, represent by their numberous combinations all our ideas and sensations! the musical characters are probably arrived at their perfection, unless emphasis, and tone, and swell could be expressed, as well as note and time. Charles the twelfth of Sweden had a design to have introduced a numeration by squares, instead of by decimation, which might have served the purposes of philosophy better than the present mode; which is said to be of Arabic invention. The alphabet is yet in a very imperfect state; perhaps seventeen letters could express all the simple sounds in the European languages. In China they have not yet learned to divide their words into syllables, and are thence necessitated to employ many thousand characters; it is said above eighty thousand. It is to be wished in this ingenious age, that the European nations would accord to reform our alphabet.

Marks the gay trill, the solemn pause inscribes,
And parts with bars the undulating tribes.
Pleased round her cane-wove throne the applauding crowd
Clap'd their rude hands, their swarthy foreheads bow'd;
With loud acclaim "a present God!" they cried,
"A present God!" rebellowing shores reply'd.—
Then peal'd at intervals with mingled swell
The echoing harp, shrill clarion, horn, and shell;
While Bards ecstatic, bending oe'r the lyre,
Struck deeper chords, and wing'd the song with fire.
Then mark'd Astronomers with keener eyes
The Moon's refulgent journey through the skies;
Watch'd the swift Comets urge their blazing cars,
And weigh'd the Sun with his revolving Stars.
High raised the Chemists their Hermetic wands,
(And changing forms obey'd their waving hands,)
From Earth's deep chambers tore her golden stores,
Or fused and harden'd her chalybeate ores.
All with bent knee from fair Papyra claim
Wove by her hands the wreath of deathless fame.
——Exulting Genius crown'd his darling child,
The young Arts clasp'd her knees, and Virtue smiled.
——So now Delany forms her mimic bowers,
Her paper foliage, and her silken flowers;
Her virgin train the tender scissars ply,
Vein the green leaf, the purple petal dye:
Round wiry stems the flaxen tendril bends,
Moss creeps below, and waxen fruit impends.
Cold Winter views amid his realms of snow
Delany's vegetable statues blow;

So now Delany. l. 153. Mrs. Delany has finished nine hundred and seventy accurate and elegant representations of different vegetables with the parts of their flowers, fructification, &c. according with the classification of Linneus, in what she terms paper-mosaic. She began this work at the age of 74, when her sight would no longer serve her to paint, in which she much excelled; between her age of 74 and 82, at which time her eyes quite failed her, she executed the curious Hortus siccus above-mentioned, which I suppose contains a greater number of plants than were ever before drawn from the life by any one person. Her method consisted in placing the leaves of each plant with the petals, and all the other parts of the flowers on coloured paper, and cutting them with sissers accurately to the natural size and form, and then pasting them on a dark ground; the effect of which is wonderful, and their accuracy less liable to fallacy than drawings. She is at this time, (1788,) in her 89th year, with all the powers of a fine understanding still unimpaired. I am informed another very ingenious lady, Mrs. North, is constructing a similar Hortus siccus, or Paper-garden; which she executes on a ground of vellum with such elegant taste and scientific accuracy, that it cannot fail to become a work of inestimable value.

Smooth's his stern brow, delays his hoary wing,
And eyes with wonder all the blooms of spring.

The gentle LAPSANA, NYMPHÆA fair,
And bright CALENDULA with golden hair,
Watch with nice eye the earth's diurnal way,
Marking her solar and sidereal day,
Her slow nutation, and her varying clime,
And trace with mimic art the march of Time;
Round his light foot a magic chain they fling,
And count the quick vibrations of his wing.—
First in its brazen cell reluctant roll'd
Bends the dark spring in many a steely fold;
On spiral brass is stretch'd the wiry thong,
Tooth urges tooth, and wheel drives wheel along;
In diamond-eyes the polish'd axles flow,
Smooth slides the hand, the ballance pants below.
Round the white circlet in relievo bold
A Serpent twines his scaly length in gold;
And brightly pencil'd on the enamel'd sphere
Live the fair trophies of the passing year.
—Here *Time*'s huge fingers grasp his giant-mace,

Lapsana, Nymphœa alba, Calendula. l. 163. And many other flowers close and open their petals at certain hours of the day; and thus constitute, what Linneus calls the Horologe, or Watch of Flora. He enumerates 46 flowers, which posses this kind of sensibility, I shall mention a few of them with their respective hours of rising and setting as Linneus terms them. He divides them first into *meteoric* flowers, which less accurately observe the hour of unfolding, but are expanded sooner or later, according to the cloudiness, moisture, or pressure of the atmosphere. 2d. *Tropical* flowers open in the morning and close before evening every day, but the hour of the expanding becomes earlier or later as the length of the day increases or decreases. 3d. *Æquinoctial* flowers, which open at a certain and exact hour of the day, and for the most part close at another determinate hour.

Hence the Horologe or watch of Flora is formed from numerous plants, of which the following are those most common in this country. Leontodon taraxacum, Dandelion, opens at 5–6, closes at 8–9. Hieracium pilosella, mouse-ear hawkweed, opens at 8, closes at 2. Sonchus lævis, smooth Sow-thistle, at 5 and at 11–12. Lactuca sativa, cultivated Lettice, at 7 and 10. Tragopogon luteum, yellow Goatsbeard, at 3–5 and at 9–10. Lapsana, nipplewort, at 5–6 and 10–1. Nymphæa alba, white water lily, at 7 and 5. Papaver nudicaule, naked poppy, at 5 and at 7. Hemerocallis fulva, tawny Day-lily, 5 and at 7–8. Convolvulus, at 5–6. Malva, Mallow, at 9–10, and at 1. Arenarea purpurea, purple Sandwort, at 9–10, and at 2–3. Anagallis, pimpernel, at 7–8. Portulaca hortensis, garden Purslain, at 9–10, and at 11–12. Dianthus prolifer, proliferous Pink, at 8, and at 1. Cichoreum. Succory, at 4–5. Hypochæris, at 6–7 and at 4–5. Crepis at 4–5, and at 10–11. Picris, at 4–5, and at 12. Calendula field, at 9, and at 3. Calendula african, at 7, and at 3–4.As these observations were probably made in the botanic gardens at Upsal, they must require further attention to suit them to our climate. See Stillingfleet's calendar of Flora.

And dash proud Superstition from her base,
Rend her strong towers and gorgeous fanes, and shed
The crumbling fragments round her guilty head.
There the gay *Hours*, whom wreaths of roses deck,
Lead their young trains amid the cumberous wreck;
And, slowly purpling o'er the mighty waste,
Plant the fair growths of Science and of Taste.
While each light *Moment*, as it dances bye
With feathery foot and pleasure-twinkling eye,
Feeds from its baby-hand, with many a kiss,
The callow nestlings of domestic Bliss.

As yon gay clouds, which canopy the skies,
Change their thin forms, and loose their lucid dyes;
So the soft bloom of Beauty's vernal charms
Fades in our eyes, and withers in our arms.
—Bright as the silvery plume, or pearly shell,
The snow-white rose, or lily's virgin bell,
The fair HELLEBORAS attractive shone,
Warm'd every Sage, and every Shepherd won.—
Round the gay sisters press the enamour'd bands,
And seek with soft sollicitude their hands.
—Ere while how changed!—in dim suffusion lies
The glance divine, that lighten'd in their eyes;
Cold are those lips, where smiles seductive hung,
And the weak accents linger on their tongue;
Each roseat feature fades to livid green,—
—Disgust with brow averted shuts the scene.

So from his gorgeous throne, which awed the world,
The mighty Monarch of the east was hurl'd,
To dwell with brutes beneath the midnight storm,
By Heaven's just vengeance changed in mind and form.
—Prone to the earth He bends his brow superb,

Helleborus. l. 199. Many males, many females. The Helleborus niger, or Christmas rose, has a large beautiful white flower, adorned with a circle of tubular two-lip'd nectaries. After impregnation the flower undergoes a remarkable change, the nectaries drop off, but the white corol remains and gradually becomes quite green. This curious metamorphose of the corol, when the nectaries fall off, seems to shew that the white juices of the corol were before carried to the nectaries, for the purpose of producing honey; because when these nectaries fall off, no more of the white juice is secreted in the corol, but it becomes green, and degenerates into a calyx. See note on Lonicera. The nectary of the Tropæolum, garden nasturtion, is a coloured horn growing from the calyx.

Crops the young floret and the bladed herb;
Lolls his red tongue, and from the reedy side
Of slow Euphrates laps the muddy tide.
Long eagle-plumes his arching neck invest,
Steal round his arms, and clasp his sharpen'd breast;
Dark brinded hairs in bristling ranks, behind,
Rise o'er his back, and rustle in the wind,
Clothe his lank sides, his shrivel'd limbs surround,
And human hands with talons print the ground.
Silent in shining troops the Courtier-throng
Pursue their monarch, as he crawls along;
E'en Beauty pleads in vain with smiles and tears,
Nor Flattery's self can pierce his pendant ears.

Two Sister-Nymphs to Ganges' flowery brink
Bend their light steps, the lucid water drink,
Wind through the dewy rice, and nodding canes,
(As *eight* black Eunuchs guard the sacred plains,)
With playful malice watch the scaly brood,
And shower the inebriate berries on the flood.—
Stay in your crystal chambers, silver tribes!
Turn your bright eyes, and shun the dangerous bribes;
The tramel'd net with less destruction sweeps
Your curling shallows, and your azure deeps;
With less deceit, the gilded fly beneath,
Lurks the fell hook unseen,—to taste is death!—
—Dim your slow eyes, and dull your pearly coat,
Drunk on the wave your languid forms shall float,
On useless fins in giddy circles play,
And Herons and Otters seize you for their prey.—

So, when the Saint from Padua's graceless land
In silent anguish sought the barren strand,
High on the shatter'd beech sublime He stood,
Still'd with his waving arm the babling flood;

Two Sister-Nymphs. l. 227. Menispermum. Cocculus. Indian berry. Two houses, twelve males. In the female flower there are two styles and eight filaments without anthers on their summits; which are called by Linneus eunuchs. See the note on Curcuma. The berry intoxicates fish. Saint Anthony of Padua, when the people refused to hear him, preached to the fish, and converted them. Addison's travels in Italy.

"To Man's dull ear," He cry'd, "I call in vain,
Hear me ye scaly tenants of the main!"—
Mishapen Seals approach in circling flocks,
In dusky mail the Tortoise climbs the rocks,
Torpedoes, Sharks, Rays, Turbots, Dolphins pour
Their twinkling squadrons round the glittering shore;
With tangled fins, behind, huge Phocæ glide,
And Whales and Grampi swell the distant tide.
Then kneel'd the hoary Seer, to heaven address'd
His fiery eyes, and smote his sounding breast;
"Bless ye the Lord!' with thundering voice he cry'd,
"Bless ye the Lord!" the bending shores reply'd;
The winds and waters caught the sacred word,
And mingling echoes shouted "Bless the Lord!"
The listening shoals the quick contagion feel,
Pant on the floods, inebriate with their zeal,
Ope their wide jaws, and bow their slimy heads,
And dash with frantic fins their foamy beds.

Sopha'd on silk, amid her charm-built towers,
Her meads of asphodel, and amaranth bowers,
Where Sleep and Silence guard the soft abodes,
In sullen apathy PAPAVER nods.
Faint o'er her couch in scintillating streams
Pass the thin forms of Fancy and of Dreams;
Froze by inchantment on the velvet ground
Fair youths and beauteous ladies glitter round;
On crystal pedestals they seem to sigh,
Bend the stiff knee, and lift the unmoving eye.
—And now the Sorceress bares her shrivel'd hand,
And circles thrice in air her ebon wand;
Fill'd with new life descending statues talk,
The pliant marble softening as they walk;

Papaver. l. 268. Poppy. Many males, many females. The plants of this class are almost all of them poisonous; the finest opium is procured by wounding the heads of large poppies with a three-edged knife, and tying muscle shells to them to catch the drops. In small quantities it exhilerates the mind, raises the passions, and invigorates the body; in large ones it is succeeded by intoxication, languor, stupor and death. It is customary in India for a messenger to travel above a hundred miles without rest or food, except an apropriated bit of opium for himself, and a larger one for his horse at certain stages. The emaciated and decrepid appearance with the ridiculous and idiotic gestures of the opium-eaters in Constantinople is well described in the memoirs of Baron de Tott.

With warmer lips relenting damsels speak,
And brighter blushes tinge the Parian cheek;
To viewless lutes aerial voices sing,
And hovering Loves are heard on rustling wing.
—She waves her wand again! —fresh horrors seize
Their stiffening limbs, their vital currents freeze;
By each cold numph her marble lover lies,
And iron slumbers seal their glassy eyes.
So with his dread Caduceus HERMES led
From the dark regions of the imprison'd dead,
Or drove in silent shoals the lingering train
To Night's dull shore, and PLUTO's dreary reign.

So with her waving pencil CREWE commands
The realms of Taste, and Fancy's fairy Lands;
Calls up with magic voice the shapes, that sleep
In earth's dark bosom, or unfathom'd deep;
That shrined in air on viewless wings aspire,
Or blazing bathe in elemental fire.
As with nice touch her plastic hand she moves,
Rise the fine forms of Beauties, Graces, Loves;
Kneel to the fair Inchantress, smile or sigh,
And fade or flourish, as she turns her eye.

Fair CISTA, rival of the rosy dawn,
Call'd her light choir, and trod the dewy lawn;

So with her waving pencil. l. 291 Alluding to the many beautiful paintings by Miss EMMA CREWE. To whom the author is indebted for the very elegant Frontispiece, where Flora at play with Cupid is loading him with garden-tools.

Cistus lubdaniferus. l. 301. Many males, one female. The petals of this beautiful and fragrant shrub, as well as of the Oenothera, tree primrose, and others, continue expanded but a few hours, falling off about noon, or soon after in hot weather. The most beautiful flower of the Cactus grandiflorus, (see Cerea) are of equally short duration, but have their existence in the night. And the flowers of the Hibiscus trionum are said to continue but a single hour. The courtship between the males and females in these flowers might be easily watched; the males are said to approach and recede from the females alternately. The flowers of the Hibiscus sinensis, mutable rose, live in the West Indies, their native climate, but one day; but have this remarkable property, they are white at the first expansion, then change to deep red, and become purple as they decay.

The gum or resin of this fragrant vegetable is collected from extensive underwoods of it in the East by a singular contrivance. Long leathern thongs are tied to poles and cords, and drawn over the tops of these shrubs about noon; which thus collect the dust of the anthers, which adheres to the leather, and is occasionally scraped off. Thus in some degree is the manner imitated, in which the bee collects on his thighs and legs the same material for the construction of his combs.

Hail'd with rude melody the new-born May,
As cradled yet in April's lap she lay.

I.

"Born in yon blaze of orient sky,
Sweet MAY! thy radiant form unfold;
Unclose thy blue voluptuous eye,
And wave thy shadowy locks of gold.

II.

For Thee the fragrant zephyrs blow,
For Thee descends the funny shower;
The rills in softer murmurs flow,
And brighter blossoms gem the bower.

III.

Light Graces dress'd in flowery wreaths
And tiptoe Joys their hands combine;
And Love his sweet contagion breathes,
And laughing dances round thy shrine.

IV.

Warm with new life the glittering throngs
On quivering fin and rustling wing
Delighted join their votive songs,
And hail thee, Goddess of the Spring."

O'er the green brinks of Severn's oozy bed,
In changeful rings, her sprightly troop She led;
PAN trip'd before, where Eudness shades the mead,

And blew with glowing lip his sevenfold reed;
Emerging Naiads swell'd the jocund strain,
And aped with mimic step the dancing train.—
"I faint, I fall!"—*at noon* the Beauty cried,
"Weep o'er my tomb, ye Nymphs!"—and sunk and died.
—Thus, when white Winter o'er the shivering clime
Drives the still snow, or showers the silver rime;
As the lone shepherd o'er the dazzling rocks
Prints his steep step, and guides his vagrant flocks;
Views the green holly veil'd in network nice,
Her vermil clusters twinkling in the ice;
Admires the lucid vales, and slumbering floods,
Fantastic cataracts, and crystal woods,
Transparent towns, with seas of milk between,
And eyes with transport the refulgent scene:—
If breaks the sunshine o'er the spangled trees,
Or flits on tepid wing the western breeze,
In liquid dews descends the transient glare,
And all the glittering pageant melts in air.

Where Andes hides his cloud-wreath'd crest in snow,
And roots his base on burning sands below;
CINCHONA, fairest of Peruvian maids,
To Health's bright Goddess in the breezy glades
On Quito's temperate plain an altar rear'd,
Trill'd the loud hymn, the solemn prayer prefer'd:
Each balmy bud she cull'd, and honey'd flower,
And hung with fragrant wreaths the sacred bower;
Each pearly sea she search'd, and sparkling mine,
And piled their treasures on the gorgeous shrine;
Her suppliant voice for sickening Loxa raised,—
Sweet breath'd the gale, and bright the censor blazed.
—"Divine HYGIEA! on thy votaries bend
Thy angel-looks, oh, hear us, and defend!

Sevenfold reed. l. 324. The sevenfold reed, with which Pan is frequently described, seems to indicate, that he was the inventor of the musical gamut.

Cinchona. l. 345. Peruvian bark-tree. Five males, and one female. Several of these trees were fell'd for other purposes into a lake, when an epidemic fever of a very mortal kind prevailed at Loxa in Peru, and the woodmen accidentally drinking the water were cured, and thus were discovered the virtues of this famous drug.

While streaming o'er the night with baleful glare
The star of Autumn rays his misty hair;
Fierce from his fens the Giant AGUE springs,
And wrap'd in fogs descends on vampire-wings;
Before, with shuddering limbs cold Tremor reels,
And Fever's burning nostril dogs his heels;
Loud claps the grinning Fiend his iron hands,
Stamps with his marble feet, and shouts along the lands;
Withers the damask cheek, unnerves the strong,
And drives with scorpion-lash the shrieking throng.
Oh, Goddess! on thy kneeling votaries bend
Thy angel-looks, oh, hear us, and defend!"
— HYGIEA, leaning from the blest abodes,
The crystal mansions of the immortal gods,
Saw the sad Nymph uplift her dewy eyes,
Spread her white arms, and breathe her fervid sighs;
Call'd to her fair associates, Youth, and Joy,
And shot all-radiant through the glittering sky;
Loose waved behind her golden train of hair,
Her sapphire mantle swam diffus'd in air.—
O'er the grey matted moss, and pansied sod,
With step sublime the glowing Goddess trod,
Gilt with her beamy eye the conscious shade,
And with her Smile celestial bless'd the maid.
"Come to my arms," with seraph voice she cries,
"Thy vows are heard, benignant Nymph! arise;
Where yon aspiring trunks fantastic wreath
Their mingled roots, and drink the rill beneath,
Yield to the biting axe thy sacred wood,
And strew the bitter foliage on the flood."
In silent homage bow'd the blushing maid,—
Five youths athletic hasten to her aid,
O'er the scar'd hills re-echoing strokes resound,
And headlong forests thunder on the ground.
Round the dark roots, rent bark, and shatter'd boughs,
From ocherous beds the swelling fountain flows;
With streams austere its winding margin laves,
And pours from vale to vale its dusky waves.
—As the pale squadrons, bending o'er the brink,
View with a sigh their alter'd forms, and drink;
Slow-ebbing life with refluent crimson breaks
O'er their wan lips, and paints their haggard cheeks;
Through each fine nerve rekindling transports dart,

Light the quick eye, and swell the exulting heart.
—Thus ISRAEL's heaven-taught chief o'er trackless sands
Led to the sultry rock his murmuring bands.
Bright o'er his brows the forky radiance blazed,
And high in air the rod divine He rais'd.—
Wide yawns the cliff!—amid the thirsty throng
Rush the redundant waves, and shine along;
With gourds and shells and helmets press the bands,
Ope their parch'd lips, and spread their eager hands,
Snatch their pale infants to the exuberant shower,
Kneel on the shatter'd rock, and bless the almighty Power.

Bolster'd with down, amid a thousand wants,
Pale Dropsy rears his bloated form, and pants;
"Quench me, ye cool pellucid rills!" he cries,
Wets his parch'd tongue, and rolls his hollow eyes.
So bends tormented TANTALUS to drink,
While from his lips the refluent waters shrink;
Again the rising stream his bosom laves,
And Thirst consumes him 'mid circumfluent waves.
—Divine HYGIEA, from the bending sky
Descending, listens to his piercing cry;
Assumes bright DIGITALIS' dress and air,
Her ruby cheek, white neck, and raven hair;
Four youths protect her from the circling throng,
And like the Nymph the Goddess steps along.—
—O'er Him She waves her serpent-wreathed wand,

Digitalis. l. 421. Of the class Two Powers. Four males, one female, Foxglove. The effect of this plant in that kind of Dropsy, which is termed anasarca, where the legs and thighs are much swelled, attended with great difficulty of breathing, is truly astonishing. In the ascites accompanied with anasarca of people past the meridian of life it will also sometimes succeed. The method of administering it requires some caution, as it is liable in greater doses to induce very violent and debilitating sickness, which continues one or two days, during which time the dropsical collection however disappears. One large spoonful, or half an ounce, of the following decoction, given twice a day, will generally succeed in a few days. But in more robust people, one large spoonful every two hours till four spoonfuls are taken, or till sickness occurs, will evacuate the dropsical swellings with greater certainty, but is liable to operate more violently. Boil four ounces of the fresh leaves of purple Foxglove (which leaves may be had at all seasons of the year) from two pints of water to twelve ounces, add to the strained liquor, while yet warm, three ounces of rectified spirit of wine. A theory of the effects of this medicine with many successful cases may be seen in a pamphlet called, "Experiments on mucilaginous and purulent matter," published by Dr. Darwin in 1780. Sold by Cadel, London.

Cheers with her voice, and raises with her hand,
Warms with rekindling bloom his visage wan,
And charms the shapeless monster into man.

So when Contagion with mephitic breath
And wither'd Famine urged the work of death;
Marseille's good Bishop, London's generous Mayor,[5]
With food and faith, with medicine and with prayer,
Raised the weak head and stay'd the parting sigh,
Or with new life relumed the swimming eye.—
—And now, BENEVOLENCE! thy rays divine
Dart round the globe from Zembla to the Line;
O'er each dark prison plays the cheering light,
Like northern lustres o'er the vault of night.—
From realm to realm, with cross or crescent crown'd,
Where'ere Mankind and Misery are found,
O'er burning sands, deep waves, or wilds of snow,
Thy HOWARD journeying seeks the house of woe.
Down many a winding step to dungeons dank,
Where anguish wails aloud, and fetters clank;
To caves bestrew'd with many a mouldering bone,
And cells, whose echoes only learn to groan;
Where no kind bars a whispering friend disclose,
No sunbeam enters, and no zephyr blows,
HE treads, inemulous of fame or wealth,
Profuse of toil, and prodigal of health;
With soft assuasive eloquence expands
Power's rigid heart, and opes his clenching hands;

Marseille's good Bishop. l. 431. In the year 1720 and 1722, the Plague made dreadful havock at Marseilles. At which times the Bishop was indefatigable in the execution of his pastoral office, visiting, relieving, encouraging, and absolving them with extream tenderness. (Geograph. Dict. printed for Coot, Lond.) and though perpetually exposed to the infection like Sir John Lawrence mentioned below, they both are said to have escaped the disease.

London's generous Mayor. l. 431. During the great Plague at London in the year 1665, Sir John Lawrence, the then Lord Mayor, continued the whole time in the city, heard complaints and redressed them, enforced the wisest regulations then known, and saw them executed. The day after the disease was known with certainty to be the Plague, above 40,000 servants were dismissed, and turned into the streets to perish; for no one would receive them into their houses; and the villages near London drove them away with pitch-forks and fire-arms. Sir John Lawrence supported them all, as well as the needy who were sick, at first by expending his own fortune, till subscriptions could be sollicited and received from all parts of the nation. *Journal of the Plague-year. Printed for E. Nut. &c. at the R. Exchange.* 1722.

Leads stern-eye'd Justice to the dark domains,
If not to sever, to relax the chains;
Or guides awaken'd Mercy through the gloom,
And shews the prison, sister to the tomb!—
Gives to her babes the self-devoted wife,
To her fond husband liberty and life!—
—The Spirits of the Good, who bend from high
Wide o'er these earthly scenes their partial eye,
When first, array'd in VIRTUE's purest robe,
They saw her HOWARD traversing the globe;
Saw round his brows her sun-like Glory blaze
In arrowy circles of unwearied rays;
Mistook a Mortal for an Angel-Guest,
And ask'd what Seraph-foot the earth imprest.
—Onward he moves!—Disease and Death retire,
And murmuring Demons hate him, and admire.

Here paused the Goddess,—on HYGIEA's shrine
Obsequious Gnomes repose the lyre divine;
Descending Sylphs relax the trembling strings,
And catch the rain-drops on their shadowy wings.
—And now her vase a modest Naiad fills
With liquid crystal from her pebbly rills;
Piles the dry cedar round her silver urn,
(Bright climbs the blaze, the crackling faggots burn,)
Culls the green herb of China's envy'd bowers,
In gaudy cups the steamy treasure pours;
And, sweetly-smiling, on her bended knee
Presents the fragrant quintessence of Tea.

INTERLUDE. II.

Bookseller. THE monsters of your Botanic Garden are as surprising as the bulls 1
with brazen feet, and the fire-breathing dragons, which guarded the Hesperian fruit; yet are they not disgusting, nor mischievous: and in the manner you have chained them together in your exhibition, they succeed each other amusingly enough, like prints of the London Cries, wraped upon rollers, with a glass before them. In this at least they resemble the monsters in Ovid's Metamorphoses; but your similies, I suppose, are Homeric?

Poet. The great Bard well understood how to make use of this kind of ornament 2
in Epic Poetry. He brings his valiant heroes into the field with much parade, and sets them a fighting with great fury; and then after a few thrusts and parries he introduces a long string of similies. During this the battle is supposed to continue; and thus the time necessary for the action is gained in our imaginations; and a degree of probability produced, which contributes to the temporary deception or reverie of the reader.

But the similies of Homer have another agreeable characteristic; they do not 3
quadrate, or go upon all fours, (as it is called) like the more formal similies of some modern writers; any one resembling feature seems to be with him a sufficient excuse for the introduction of this kind of digression; he then proceeds to deliver some agreeable poetry on this new subject, and thus converts every simile into a kind of short episode.

B. Then a similie should not very accurately resemble the subject? 4

P. No, it would then become a philosophical analogy, it would be ratiocination 5
instead of poetry: it need only so far resemble the subject, as poetry itself ought to resemble nature. It should have so much sublimity, beauty, or novelty, as to interest the reader; and should be expressed in picturesque language, so as to bring the scenery before his eye; and should lastly bear so much veri-similitude as not to awaken him by the violence of improbability or incongruity.

B. May not the reverie of the reader be dissipated or disturbed by disagreeable 6
images being presented to his imagination, as well as by improbable or incongruous ones?

P. Certainly, he will endeavour to rouse himself from a disagreeable reverie, as from the night-mare. And from this may be discovered the line of boundary between the Tragic and the Horrid: which line however will veer a little this way or that according to the prevailing manners of the age or country, and the peculiar associations of ideas, or idiosyncracy of mind, of individuals. For instance, if an artist should represent the death of an officer in battle by shewing a little blood on the bosom of his shirt, as if a bullet had there penetrated, the dying figure would affect the beholder with pity; and if fortitude was at the same time expressed in his countenance, admiration would be added to our pity. On the contrary, if the artist should chuse to represent his thigh as shot away by a cannon ball, and should exhibit the bleeding flesh and shattered bone of the stump, the picture would introduce into our minds ideas from a butcher's shop, or a surgeon's operation-room, and we should turn from it with disgust. So if characters were brought upon the stage with their limbs disjointed by torturing instruments, and the floor covered with clotted blood and scattered brains, our theatric reverie would be destroyed by disgust, and we should leave the playhouse with detestation. 7

The Painters have been more guilty in this respect than the Poets; the cruelty of Apollo in flaying Marcias alive is a favorite subject with the antient artists; and the tortures of expiring martyrs have disgraced the modern ones. It requires little genius to exhibit the muscles in convulsive action either by the pencil or the chissel, because the interstices are deep, and the lines strongly defined: but those tender gradations of muscular action, which constitute the graceful attitudes of the body, are difficult to conceive or to execute, except by a master of nice discernment and cultivated taste. 8

B. By what definition would you distinguish the Horrid from the Tragic? 9

P. I suppose the latter consists of Distress attended with Pity, which is said to be allied to Love, the most agreeable of all our passions: and the former in Distress accompanied with Disgust, which is allied to Hate, and is one of our most disagreeable sensations. Hence when horrid scenes of cruelty are represented in pictures, we wish to disbelieve their existence, and voluntarily exert ourselves to escape from the deception: whereas the bitter cup of true Tragedy is mingled with some sweet consolatory drops, which endear our tears, and we continue to contemplate the interesting delusion with a delight, which it is not easy to explain. 10

B. Has not this been explained by Lucretius, where he describes a shipwreck; and says, the Spectators receive pleasure from feeling themselves safe on land? and by Akinside in his beautiful poem on the Pleasures of Imagination, who ascribes it to our finding objects for the due exertion of our passions? 11

P. We must not confound our sensations at the contemplation of real misery with those, which we experience at the scenical representations of tragedy. The spectators of a shipwreck may be attracted by the dignity and novelty of the object; and from these may be said to receive pleasure, but not from the distress of the sufferers. But at the exhibition of a good tragedy, we are not only amused by the dignity and novelty and beauty of the objects before us; but, if any distressful 12 circumstances occur too forceable for our sensibility, we can voluntarily exert ourselves, and recollect, that the scenery is not real: and thus not only the pain, which we had received from the apparent distress, is lessened; but a new source of pleasure is opened to us, similar to that which we frequently have felt on awaking from a distressful dream; we are glad that it is not true. We are at the same time unwilling to relinquish the pleasure, which we receive from the other interesting circumstances of the drama; and on that account quickly permit ourselves to relapse into the delusion; and thus alternately believe and disbelieve, almost every moment, the existence of the objects represented before us.

B. Have those two Sovereigns of poetic land, HOMER and SHAKESPEAR, kept their 13
works intirely free from the Horrid?—or even yourself in your third Canto?

P. The descriptions of the mangled carcasses of the companions of Ulysses, 14
in the cave of Polypheme, is in this respect certainly objectionable, as is well obverved by Scaliger. And in the play of Titus Andronicus, if that was written by Shakespear (which from its internal evidence I think very improbable,) there are many horrid and disgustful circumstances. The following Canto is submitted to the candour of the critical reader, to whose opinion I shall submit in silence.

THE LOVES OF THE PLANTS. CANTO III

AND now the Goddess sounds her silver shell,
And shakes with deeper tones the inchanted dell;
Pale, round her grassy throne, bedew'd with tears,
Flit the thin forms of Sorrows, and of Fears;
Soft Sighs responsive whisper to the chords,
And Indignations half-unsheath their swords.

Thrice round the grave CIRCÆA prints her tread,
And chaunts the numbers, which disturb the dead;
Shakes o'er the holy earth her sable plume,
Waves her dread wand, and strikes the echoing tomb!
—Pale shoot the stars across the troubled night,
The timorous moon withholds her conscious light;

Circæa. l. 7. Enchanter's Nightshade. Two males, one female. It was much celebrated in the mysteries of witchcraft, and for the purpose of raising the devil, as its name imports. It grows amid the mouldering bones, and decayed coffins, in the ruinous vaults of Sleaford-church in Lincolnshire. The superstitious ceremonies or histories belonging to some vegetables have been truly ridiculous; thus the Druids are said to have cropped the misletoe with a golden axe or sickle; and the Bryony or Mandrake, was said to utter a scream, when its root was drawn from the ground; and that the animal which drew it up, became diseased and soon died; on which account, when it was wanted for the purposes of medicine, it was usual to loosen and remove the earth about the root, and then to tie it by means of a cord to a dog's tail; who was whiped to pull it up, and was then supposed to suffer for the impiety of the action. And even at this day bits of dried root of Peony are rubbed smooth, and strung, and sold under the name of Anodyne necklaces, and tied round the necks of children, to facilitate the growth of their teeth! add to this that in Price's History of Cornwall, a book published about ten years ago, the Virga divinatoria, or divining rod, has a degree of credit given to it. This rod is of hazle, or other light wood, and held horizontally in the hand, and is said to bow towards the ore, whenever the Conjurer walks over a mine. A very few years ago in France, and even in England, another kind of divining rod has been used to discover springs of water in a similar manner, and gained some credit. And in the very last year, there were many in France and some in England, who underwent an enchantment without any divining rod at all, and believed themselves to be affected by an invisible agent; which the Enchanter called Animal Magnatism!

Shrill scream the famish'd bats, and shivering owls,
And loud and long the dog of midnight howls!—
—Then yawns the bursting ground!—*two* imps obscene
Rise on broad wings, and hail the baleful queen;
Each with dire grin salutes the potent wand,
And leads the sorceress with his sooty hand;
Onward they glide, where sheds the sickly yew
O'er many a mouldering bone its nightly dew;
The ponderous portals of the church unbar,—
Hoarse on their hinge the ponderous portals jar;
As through the colour'd glass the moon beam falls,
Huge shapeless spectres quiver on the walls;
Low murmurs creep along the hollow ground,
And to each step the pealing ailes resound;
By glimmering lamps, protecting saints among,
The shrines all trembling as they pass along,
O'er the still choir with hideous laugh they move,
(Fiends yell below, and angels weep above!)
Their impious march to God's high altar bend,
With feet impure the sacred steps ascend;
With wine unbless'd the holy chalice stain,
Assume the mitre, and the cope profane;
To heaven their eyes in mock devotion throw,
And to the cross with horrid mummery bow;
Adjure by mimic rites the powers above,
And plight alternate their Satanic love.

Avaunt, ye Vulgar! from her sacred groves
With maniac step the Pythian LAURA moves;
Full of the God her laboring bosom sighs,
Foam on her lips, and fury in her eyes,

Laura. l. 40. Prunus. Laura-cerasus. Twenty males, one female. The Pythian priestess is supposed to have been made drunk with infusion of laurel-leaves, when she delivered her oracles. The intoxication or inspiration is finely described by Virgil. Æn. L. vi. The distiled water from laurel leaves is perhaps the most sudden poison, we are acquainted with in this country. I have seen about two spoonfuls of it destroy a large pointer dog, in less than ten minutes. In a smaller dose it is said to produce intoxication; on this account there is reason to believe it acts in the same manner as opium and vinous spirit; but that the dose is not so well ascertained. See note on Tremella. It is used in the Ratafie of the distillers, by which some dram-drinkers have been suddenly killed. One pint of water, distiled from fourteen pounds of black cherry stones bruised, has the same deleterious effect, destroying as suddenly as Laurel-water. It is probable Apricot-kernels, Peach-leaves, Wallnut-leaves, and whatever possesses the kernel-flavour may have similar qualities.

Strong writhe her limbs, her wild dishevel'd hair
Starts from her laurel-wreath, and swims in air.—
While *twenty* Priests the gorgeous shrine surround
Cincture'd with ephods, and with garlands crown'd,
Contending hosts and trembling nations wait
The firm immutable behests of Fate;
—She speaks in thunder from her golden throne
With words *unwill'd*, and wisdom not her own.

So on his NIGHTMARE through the evening fog
Flits the squab Fiend o'er fen, and lake, and bog;
Seeks some love-wilder'd Maid with sleep oppress'd,
Alights, and grinning sits upon her breast.
—Such as of late amid the murky sky
Was mark'd by FUSSELI's poetic eye;
Whose daring tints, with SHAKESPEAR's happiest grace,
Gave to the airy phantom form and place.—
Back o'er her pillow sinks her blushing head,
Her snow-white limbs hang helpless from the bed;
While with quick sighs, and suffocative breath,
Her interrupted heart-pulse swims in death.
—Then shrieks of captured towns, and widow's tears,
Pale lovers stretch'd upon their blood-stain'd biers,
The headlong precipice that thwarts her flight,
The trackless desert, the cold starless night,
And stern-eye'd Murderer with his knife behind,
In dread succession agonize her mind.
O'er her fair limbs convulsive tremors fleet,
Start in her hands, and struggle in her feet;
In vain to scream with quivering lips she tries,
And strains in palsy'd lids her tremulous eyes;
In vain she *wills* to walk, swim, run, fly, leap;
The WILL presides not in the bower of SLEEP.

The Will presides not. l. 74. Sleep consists in the abolition of all voluntary power, both over our muscular motions and our ideas; for we neither walk nor reason in sleep. But at the same time, many of our muscular motions, and many of our ideas continue to be excited into action, in consequence of internal irritations, and of internal sensations; for the heart and arteries continue to beat, and we experience variety of passions, and even hunger and thirst in our dreams. Hence I conclude, that our nerves of sense are not torpid or inert during sleep; but that they are only precluded from the perception of external objects, by their external organs being rendered unfit to transmit to them the appulses of external bodies, during the suspension of the power of volition; thus the eyelids are closed in sleep, and I suppose the tympanum of the ear is not stretched, because they are deprived

—On her fair bosom sits the Demon-Ape
Erect, and balances his bloated shape;
Rolls in their marble orbs his Gorgon-eyes,
And drinks with leathern ears her tender cries.

Arm'd with her ivory beak, and talon-hands,
Descending FICA dives into the sands;
Chamber'd in earth with cold oblivion lies,
Nor heeds, ye Suitor-Train, your amorous sighs;
Erewhile with renovated beauty blooms,
Mounts into air, and moves her leafy plumes.
—Where HAMPS and MANIFOLD, their cliffs among,
Each in his flinty channel winds along;
With lucid lines the dusky Moor divides,
Hurrying to intermix their sister tides,
Where still their silver-bosom'd Nymphs abhor
The blood-smear'd mansion of gigantic THOR,—

---

of the voluntary exertions of the muscles appropriated to these purposes; and it is probable something similar happens to the external apparatus of our other organs of sense, which may render them unfit for their office of perception during sleep: for milk put into the mouths of sleeping babes occasions them to swallow and suck; and, if the eye-lid is a little opened in the day-light, by the exertions of disturbed sleep, the person dreams of being much dazzled. See the first Interlude.

When there arises in sleep, a painful desire to exert the voluntary motions, it is called the Night-mare, or Incubus.—When the sleep becomes so imperfect, that some muscular motions obey this exertion of desire, people have walked about, and even performed some domestic offices in sleep; one of these sleep-walkers I have frequently seen; once she smelt of a tube-rose, and sung, and drank a dish of tea in this state; her awaking was always attended with prodigious surprize and even fear; this disease had daily periods, and seemed to be of the epileptic kind.

Ficus indica. l. 80. Indian Fig-tree. Of the class Polygamy. This large tree rises with opposite branches on all sides, with long egged leaves; each branch emits a slender flexile depending appendage from its summit like a cord, which roots into the earth and rises again. Sloan. Hist. of Jamaica. Lin. Spec. Plant. See Capri-ficus.

Gigantic Thor. l. 90. Near the village of Wetton, a mile or two above Dove-Dale, near Ashburn in Derbyshire, there is a spacious cavern about the middle of the ascent of the mountain, which still retains the name of Thor's house; below is an extensive and romantic common, where the rivers Hamps and Manifold sink into the earth, and rise again in Ilam gardens, the seat of John Port, Esq. about three miles below. Where these rivers rise again, there are impressions resembling Fish, which appear to be of Jasper bedded in Limestone. Calcareous Spars, Shells converted into a kind of Agate, corallines in Marble, ores of Lead, Copper, and Zinc, and many strata of Flint, or Chert, and of Toadstone, or Lava, abound in this part of the country. The Druids are said to have offered human sacrifices inclosed in wicker idols to Thor. Thursday had its name from this Deity.

The broken appearance of the surface of many parts of this country; with the Swallows, as they are called, or basons on some of the mountains, like volcanic Craters, where the rain water sinks into the earth; and the numerous large stones, which seem to have been thrown over the land by volcanic explosions; as well as the great masses of Toadstone or Lava; evince the existence of violent

—Erst, fires volcanic in the marble womb
Of cloud-wrap'd WETTON raised the massy dome;
Rocks rear'd on rocks in huge disjointed piles
Form the tall turrets, and the lengthen'd ailes;
Broad ponderous piers sustain the roof, and wide
Branch the vast rain-bow ribs from side to side.
While from above descends in milky streams
One scanty pencil of illusive beams,
Suspended crags, and gaping gulphs illumes,
And gilds the horrors of the deepen'd glooms.
—Here oft the Naiads, as they chanced to play
Near the dread fane on THOR's returning day,
Saw from red altars streams of guiltless blood
Stain their green reed-beds, and pollute their flood;
Heard dying babes in wicker prisons wail,
And shrieks of matrons thrill the affrighted Gale;
While from dark caves infernal Echoes mock,
And Fiends triumphant shout from every rock!—
—*So* still the Nymphs emerging lift in air
Their snow-white shoulders and their azure hair;
Sail with sweet grace the dimpling streams along,
Listening the Shepherd's or the Miner's song;
But, when afar they view the Giant Cave,
On timorous fins they circle on the wave,
With streaming eyes and throbbing hearts recoil,
Plunge their fair forms, and dive beneath the soil.—
Closed round their heads reluctant eddies sink,
And wider rings successive dash the brink.—
Three thousand steps in sparry clefts they stray,
Or seek through sullen mines their gloomy way;
On beds of Lava sleep in coral cells,
Or sigh o'er jasper fish, and agate shells.
Till, where famed ILAM leads his boiling floods
Through flowery meadows and impending woods,
Pleased with light spring they leave the dreary night,

earthquakes at some early period of the world. At this time the channels of these subterraneous rivers seem to have been formed, when a long tract of rocks were raised by the sea flowing in upon the central fires, and thus producing an irresistable explosion of steam; and when these rocks again subsided, their parts did not exactly correspond, but left a long cavity arched over in this operation of nature. The cavities at Castleton and Buxton in Derbyshire, seem to have had a similar origin, as well as this cavern termed Thor's house. See Mr. Whitehurst's and Dr. Hutton's Theories of the Earth.

Impatiens. l. 131. Touch me not. The seed vessel consists of one cell with five divisions; each of these, when the seed is ripe, on being touched, suddenly folds itself into a spiral form, leaps from

And 'mid circumfluent surges rise to light;
Shake their bright locks, the widening vale pursue,
Their sea-green mantles fringed with pearly dew;
In playful groups by towering THORP they move,
Bound o'er the foaming wears, and rush into the DOVE.

With fierce distracted eye IMPATIENS stands,
Swells her pale cheeks, and brandishes her hands,
With rage and hate the astonish'd groves alarms,
And hurls her infants from her frantic arms.
—So when MEDÆA left her native soil
Unaw'd by danger, unsubdued by toil;
Her weeping fire and beckoning friends withstood,
And launch'd enamour'd on the boiling flood;
One ruddy boy her gentle lips caress'd,
And one fair girl was pillow'd on her breast;
While high in air the golden treasure burns,
And Love and Glory guide the prow by turns.
But, when Thessalia's inauspicious plain
Received the matron-heroine from the main;
While horns of triumph sound, and altars burn,
And shouting nations hail their Chief's return;
Agast, She saw new-deck'd the nuptial bed,
And proud CREUSA to the temple led;
Saw her in JASON's mercenary arms
Deride her virtues, and insult her charms;

---

the stalk, and disperses the seeds to a great distance by its elasticity. The capsule of the geranium and the beard of wild oats are twisted for a similar purpose, and dislodge their seeds on wet days, when the ground is best fitted to receive them. Hence one of these with its adhering capsule or beard, fixed on a stand, serves the purpose of an hygrometer, twisting itself more or less according to the moisture of the air.

The awn of barley is furnished with stiff points, which like the teeth of a saw are all turned towards the point of it; as this long awn lies upon the ground, it extends itself in the moist air of night, and pushes forwards the barley corn, which it adheres to; in the day it shortens as it drys, and as these points prevent it from receding, it draws up its pointed end; and thus creeping like a worm will travel many feet from the parent stem. That very ingenious Mechanic Philosopher Mr. Edgeworth once made on this principle a wooden automaton; its back consisted of soft Fir-wood, about an inch square and four feet long, made of pieces cut the cross way in respect to the fibres of the wood, and glued together; it had two feet before, and two behind, which supported the back horizontally; but were placed with their extremities, which were armed with sharp points of iron, bending backwards. Hence in moist weather the back lengthened and the two foremost feet were pushed forwards; in dry weather the hinder feet were drawn after; as the obliquity of the points of the feet prevented it from receeding. And thus in a month or two it walked across the room, which it inhabited. Might not this machine be applied as an Hygrometer to some meteorological purpose?

Saw her dear babes from fame and empire torn,
In foreign realms deserted and forlorn;
Her love rejected, and her vengeance braved,
By Him her beauties won, her virtues saved.—
With stern regard she eyed the traitor-king,
And felt, Ingratitude! thy keenest sting;
"Nor Heaven," She cried, "nor Earth, nor Hell can hold
"A Heart abandon'd to the thirst of Gold!"
Stamp'd with wild foot, and shood her horrent brow,
And call'd the furies from their dens below.
—Slow out of earth, before the festive crouds,
On wheels of fire, amid a night of clouds,
Drawn by fierce fiends arose a magic car,
Received the Queen, and hovering flamed in air.—
As with raised hands the suppliant traitors kneel,
And fear the vengeance they deserve to feel,
Thrice with parch'd lips her guiltless babes she press'd,
And thrice she clasp'd them to her tortur'd breast;
Awhile with white uplifted eyes she stood,
Then plunged her trembling poynards in their blood.
"Go, kiss your sire! go, share the bridal mirth!"
She cry'd, and hurl'd their quivering limbs on earth.
—Rebellowing thunders rock the marble towers,
And red-tongued lightnings shoot their arrowy showers;
Earth yawns!—the crashing ruin sinks!—o'er all
Death with black hands extends his mighty Pall;
Their mingling gore the Fiends of Vengeance quaff,
And Hell receives them with convulsive laugh.

Round the vext isles where fierce tornados roar,
Or tropic breezes sooth the sultry shore;
What time the eve her gauze pellucid spreads
O'er the dim flowers, and veils the misty meads;
Slow, o'er the twilight sands or leafy walks
With gloomy dignity DICTAMNA stalks;

Dictamnus. l. 184. Fraxinella. In the still evenings of dry seasons, this plant emits an inflammable air or gas, and flashes on the approach of a candle. There are instances of human creatures who have taken fire spontaneously, and been totally consumed. Phil. trans.

The odours of many flowers so delightful to our sense of smell, as well as the disagreeable scents of others, are owing to the exhalation of their essential oils. These essential oils have greater or less volatility, and are all inflammable; many of them are poisons to us, as these of Laurel and Tobacco; others posses a narcotic quality; as is evinced by the oil of cloves instantly

In sulphurous eddies round the weird dame
Plays the light gas, or kindles into flame.
If rests the traveller his weary head,
Grim MANCINELLA haunts the mossy bed,
Brews her black hebenon, and stealing near
Pours the curst venom in his tortured ear.—
Wide o'er the mad'ning throng URTICA flings
Her barbed shafts, and darts her poison'd stings.

---

relieving slight tooth-achs; from oil of cinnamon relieving the hiccup; and balsam of peru relieving the pain of some ulcers. They are all deleterious to certain insects, and hence their use in the vegetable economy, being produced in flowers or leaves to protect them from the depredations of their voracious enemies. One of the essential oils, that of turpentine, is recommended by M. de Thosse for the purpose of destroying insects, which infect both vegetables and animals. Having observed that the trees were attack'd by multitudes of small insects of different colours, (pucins ou pucerons) which injured and destroyed their young branches; he destroyed them all intirely in the following manner, he put into a bowl a few handfuls of earth, on which he poured a small quantity of oil of turpentine, he then beat the whole together with a spatula, pouring on it water till it became of the consistence of soup; with this mixture he moistened the ends of the branches, and both the insects and their eggs were destroyed, and other insects kept aloof by the scent of the turpentine. He adds that he destroyed the fleas of his puppies by once bathing them in warm water impregnated with oil of turpentine. Mem. d'Agriculture, An. 1787, Trimest. Printemp. p. 109. The shops of medicine are supplied with resins, balsams, and essential oils; and the tar and pitch for mechanical purposes are produced from these vegetable secretions.

Mancinella. l. 188. Hyppomane. With the milky juice of this tree the Indians poison their arrows; the dew-drops which fall from it, are so caustic as to blister the skin, and produce dangerous ulcers; whence many have found their death by sleeping under its shade. Variety of noxious plants abound in all countries, in our own the deadly nightshade, henbane, houndstongue, and many others are seen in almost every high-road untouched by animals. Some have asked what is the use of such abundance of poisons? the nauseous or pungent juices of some vegetables, like the thorns of others, are given them for their defence from the depredations of animals; hence the thorny plants are in general wholesome, and agreeable food to graminivorous animals. See note on Ilex. The flowers or petals of plants are perhaps in general more acrid than their leaves, hence they are much seldomer eaten by insects; this seems to have been the use of the essential oil in the vegetable economy, as observed above in the notes on the Dictamnus and on Ilex. The fragrance of plants is thus a part of their defence. These pungent or nauseous juices of vegetables have supplied the science of medicine with its principal materials. Such as purge, vomit, intoxicate, &c.

Urtica. l. 191. Nettle. The sting has a bag at its base, and a perforation near its point, exactly like the stings of wasps, and the teeth of adders; Hook. microgr. p. 142. Is the fluid contained in this bag, and pressed through the preforation into the wound made by the point, a caustic essential oil, or a concentrated vegetable acid? The vegetable poisons like the animal ones produce more sudden and dangerous effects, when instilled into a wound than when taken into the stomach; whence the families of Marsi and Psilli in antient Rome sucked the poison without injury out of wounds made by vipers; and were supposed to be indued with supernatural powers for this purpose. By the experiments related by Beocaria it appears, that four or five times the quantity taken by the mouth had about equal effects with that infused into a wound. The male flowers of the nettle are separate from the female, and the anthers are seen in fair weather to burst with force and to discharge a dust, which hovers about the plant like a cloud.

And fell Lobelia with contagious breath
Infects the light, and wings the gale with death.
—With fear and hate they blast the affrighted groves,
Yet own with tender care their kindred loves!—

So, where Palmira 'mid her wasted plains,
Her shatter'd aqueducts, and prostrate fanes,
(As the bright orb of breezy midnight pours
Long threads of silver through her gaping towers,
O'er mouldering tombs, and tottering columns gleams,
And frosts her deserts with diffusive beams,)
Sad o'er the mighty wreck in silence bends,
Lifts her wet eyes, her tremulous hands extends.—
If from lone cliffs a bursting rill expands
Its transient course, and sinks into the sands;
O'er the moist rock the fell Hyæna prowls,
The Leopard hisses, and the Panther growls;
On quivering wing the famish'd Vulture screams,
Dips his dry beak, and sweeps the gushing streams;
With foamy jaws, beneath, and sanguine tongue
Laps the lean Wolf, and pants, and runs along;
Stern stalks the Lion, on the rustling brinks
Hears the dread Snake, and trembles as he drinks;
Quick darts the scaly Monster o'er the plain,
Fold after fold, his undulating train;
And, bending o'er the lake his crested brow,
Starts at the Crocodile, that gapes below.

Lobelia. l. 193. Longiflora. Grows in the West Indies, and spreads such deleterious exhalations around it; that an oppression of the breast is felt on approaching it at many feet distance, when placed in the corner of a room or hot-house. Ingenhouz exper. on air, p. 146. Jacquini hort botanic. Vindeb. The exhalations from ripe fruit, or withering leaves, are proved much to injure the air in which they are confined; and it is probable all those vegetables which emit a strong scent may do this in a greater or less degree, from the Rose to the Lobelia. Whence the unwholesomeness in living perpetually in such an atmosphere of perfume, as some people wear about their hair, or carry in their handkerchiefs. Either Boerhave or Dr. Mead have affirmed they were acquainted with a poisonous fluid, whose vapour would presently destroy the person, who sat near it. And it is well known that the gas from fermenting liquors, or obtained from limestone, will destroy animals immersed in it; as well as the vapour of the Grotto del Cani near Naples.

So, where Palmira. l. 197. Among the ruins of Palmira, which are dispersed not only over the plains but even in the deserts, there is one single colonade above 2600 yards long, the bases of the Corinthian columns of which exceed the height of a man. And yet this row is only a small part of the remains of that one edifice! Volney's travels.

Where seas of glass with gay reflections smile
Round the green coasts of Java's palmy isle;
A spacious plain extends its upland scene,
Rocks rise on rocks, and fountains gush between;
Soft breathes the breeze, eternal summers reign,
And showers prolific bless the soil,—in vain!
—No spicy nutmeg scents the vernal gales,
Nor towering plantain shades the mid-day vales;
No grassy mantle hides the sable hills,
No flowery chaplet crowns the trickling rills;
Nor tufted moss, nor leathery lichen creeps
In russet tapestry o'er the crumbing steeps.
—No step retreating, on the sand impress'd,
Invites the visit of a second guest;
No refluent fin the unpeopled stream divides,
No revolant pinion cleaves the airy tides;
Nor handed moles, nor beaked worms return,
That mining pass the irremeable bourn.—
Fierce in dread silence on the blasted heath
Fell UPAS sits, the HYDRA-TREE of death.
Lo! from one root, the envenom'd soil below,
A thousand vegetative serpents grow;
In shining rays the scaly monster spreads
O'er ten square leagues his far-diverging heads;
Or in one trunk entwists his tangled form,
Looks o'er the clouds, and hisses in the storm.
Steep'd in fell poison, as his sharp teeth part,
A thousand tongues in quick vibration dart;
Snatch the proud Eagle towering o'er the heath,
Or pounce the lion, as he stalks beneath;
Or strew, as marshal'd hosts contend in vain,

Upas. l. 238. There is a poison-tree in the island of Java, which is said by its effluvia to have depopulated the country for 12 to 14 miles round the place of its growth. It is called in the Malayan language Bohon-Upas; with the juice of it the most poisonous arrows are prepared; and to gain this the condemned criminals are sent to the tree with proper direction both to get the juice, and to secure themselves from the malignant exhalations of the tree; and are pardoned if they bring back a certain quantity of the poison. But by the registers there kept, not one in four are said to return. Not only animals of all kinds both quadrupeds, fish, and birds, but all kinds of vegetables also are destroyed by the effluvia of the noxious tree, so that in a district of 12 or 14 miles round it, the face of the earth is quite barren, and rocky, intermixed only with the skeletons of men and animals; affording a scene of melancholy beyond what poets have described, or painters delineated. Two younger trees of its own species are said to grow near it. See London Magazine for 1784, or 1783. Translated from a description of the poison tree of the island of Java, written in Dutch by N. P. Foereh. For a further account of it see a note at the end of the work.

With human skeletons the whiten'd plain.
—Chain'd at his root two infant Dæmons dwell,
Breathe the soft hiss, or try the tender yell;
Rise, fluttering in the air on callow wings,
And aim at insect-prey their little stings.
So Time's strong arms with sweeping scythe erase
Art's cumberous works, and empires, from their base;
While each young Hour its sickle fine employs,
And crops the sweet buds of domestic joys!

Two Harlot-Nymphs, the fair CUSCUTAS, please
With labour'd negligence, and studied ease;
In the meek garb of modest worth disguised,
The eye averted, and the smile chastised,
With sly approach they spread their dangerous charms,
And round their victim wind their wiry arms.
So by Scamander when LAOCOON stood,
Where Troy's proud turrets glitter'd in the flood,
Raised high his arm, and with prophetic call
To shrinking realms announced her fated fall;
Whirl'd his fierce spear with more than mortal force,
And pierced the thick ribs of the echoing horse;

Cuscuta. l. 259. Dodder. Four males, 2 females. This parasite plant, (the seed splitting without cotyledons,) protrudes a spiral body, and not endeavouring to root itself in the earth ascends the vegetables in its vicinity, spirally W. S. E. or contrary to the movement of the sun; and absorbs its nourishment by vessels apparently inserted into its supporters. It bears no leaves except here and there a scale very small membranous and close under the branch. Lin. Spec. plant. edit a Reichard. Vol. I. p. 352. The Rev. T. Martyn in his elegant letters on botany adds, that not content with support, where it lays hold, there it draws its nourishment; and at length in gratitude for all this strangles its entertainer. Let. xv. A contest for air and light obtains throughout the whole vegetable world; shrubs rise above herbs; and by precluding the air and light from them, injure or destroy them; trees suffocate or incommode shrubs; the parasite climbing plants, as Ivy, Clematis, incommode the taller trees; and other parasites, which exist without having roots on the ground, as Misletoe, Tillandsia, and the mosses and funguses, incommode them all.

Some of the plants with voluble stems ascend other plants spirally east-south-west, as Humulus, Hop, Lonicera, Honey-suckle, Tamus, black Bryony, Helxine. Others turn their spiral stems W.S.E. as Convolvulus, Corn-bind, Phaseolus, Kidney-bean, Basella, Cynanche, Euphorbia, Eupatorium. The proximate or final causes of this difference have not been investigated. One of the Indian grasses, Panicum arborescens, whose stem is no thicker than a goose-quill, rises as high as the tallest trees in this contest for light and air. Spec. plant. a Reichard, V. I. p. 161. The tops of many climbing plants are tender from their quick growth; and, when deprived of their acrimony by boiling, are an agreeable article of food. The Hop-tops are in common use; I have eaten the tops of white Bryony, Bryonia alba, and found them nearly as grateful as Asparagus, and think this plant might be profitably cultivated as an early garden-vegetable. The Tamus, (called black Bryony) was less agreeable to the taste when boiled. See Galanthus.

Two Serpent-forms incumbent on the main,
Lashing the white waves with redundant train,
Arch'd their blue necks, and shook their towering crests,
And plough'd their foamy way with speckled breasts;
Then darting fierce amid the affrighted throngs,
Roll'd their red eyes, and shot their forked tongues.—
—Two daring Youths to guard the hoary fire
Thwart their dread progress, and provoke their ire.
Round sire and sons the scaly monsters roll'd,
Ring above ring, in many a tangled fold,
Close and more close their writhing limbs surround,
And fix with foamy teeth the invenom'd wound.
—With brow upturn'd to heaven the holy Sage
In silent agony sustains their rage;
While each fond Youth, in vain, with piercing cries
Bends on the tortured Sire his dying eyes.

"Drink deep, sweet youths," seductive Vitis cries,
The maudlin tear-drop glittering in her eyes;
Green leaves and purple clusters crown her head,
And the tall Thyrsus stays her tottering tread.
—*Five* hapless swains with soft assuasive smiles
The harlot meshes in her deathful toils;
"Drink deep," she carols, as she waves in air
The mantling goblet, "and forget your care."—
O'er the dread feast malignant Chemia scowls,
And mingles poison in the nectar'd bowls;
Fell Gout peeps grinning through the flimsy scene,
And bloated Dropsy pants behind unseen;
Wrap'd in his robe white Lepra hides his stains,
And silent Frenzy writhing bites his chains.

Vitis. l. 287. Vine. Five males one female. The juice of the ripe grape is a nutritive and agreeable food, consisting chiefly of sugar and mucilage. The chemical process of fermentation converts this sugar into spirit, converts food into poison! And it has thus become the curse of the christian world, producing more than half of our chronical diseases; which Mahomet observed, and forbad the use of it to his disciples. The Arabians invented distillation; and thus, by obtaining the spirit of fermented liquors in a less diluted state, added to its destructive quality. A theory of the Diabætes and Dropsy produced by drinking fermented or spirituous liquors, is explained in a treatise on the inverted motions of the lymphatic system, published by Dr. Darwin. Cadell.

So when PROMETHEUS, fearless of his ire,
Stole from the throne of JOVE forbidden fire;
And, lantern'd in his breast, from realms of day,
Bore the bright treasure to his Man of clay;—
High on cold Caucasus by VULCAN bound,
The lean impatient Vulture fluttering round,
His writhing limbs in vain he twists and strains
To break or loose the adamantine chains.
The gluttonous bird, regardless of his pangs,
Tears his swoln liver with remorseless fangs.

The gentle CYCLAMEN with dewy eye
Breathes o'er her lifeless babe the parting sigh;
And, bending low to earth, with pious hands
Inhumes her dear Departed in the sands.
"Sweet Nursling! withering in thy tender hour,
"Oh, sleep," She cries, "and rise a fairer flower!"
—So when the Plague o'er London's gasping crouds
Shook her dank wing, and steer'd her murky clouds;
When o'er the friendless bier no rites were read,
No dirge slow-chanted, and no pall out-spread;
While Death and Night piled up the naked throng,
And Silence drove their ebon cars along;

Prometheus. l. 301. The antient story of Prometheus, who concealed in his bosom the fire he had stolen, and afterwards had a vulture perpetually gnawing his liver, affords so apt an allegory for the effects of drinking spirituous liquors, that one should be induced to think the art of distillation, as well as some other chemical processes, (such as calcining gold,) had been known in times of great antiquity, and lost again. The swallowing drams cannot be better represented in hieroglyphic language than by taking fire into ones bosom; and certain it is, that the general effect of drinking fermented or spirituous liquors is an inflamed, schirrous, or paralytic liver, with its various critical or consequential diseases, as leprous eruptions on the face, gout, dropsy, epilepsy, insanity. It is remarkable that all the diseases from drinking spirituous or fermented liquors are liable to become hereditary, even to the third generation; gradually increasing, if the cause be continued, till the family becomes extinct.

Cyclamen. l. 311. Shew-bread, or Sow-bread. When the seeds are ripe, the stalk of the flower gradually twists itself spirally downwards, till it touches the ground, and forcibly penetrating the earth lodges its seeds; which are thought to receive nourishment from the parent root, as they are said not to be made to grow in any other situation.

The Trifolium subterraneum, subterraneous trefoil, is another plant, which buries its seed, the globular head of the seed penetrating the earth; which however in this plant may be only an attempt to conceal its seeds from the ravages of birds; for there is another trefoil, the trifolium globosum, or globular woolly-headed trefoil, which has a curious manner of concealing its seeds; the lower florets only have corols and are fertile; the upper ones wither into a kind of wool, and forming a head compleatly conceal the fertile calyxes. Lin. Spec. Plant. a Reichard.

Six lovely daughters, and their father, swept
To the throng'd grave CLEONE saw, and wept;
Her tender mind, with meek Religion fraught,
Drank all-resigned Affliction's bitter draught;
Alive and listening to the whisper'd groan
Of other's woes, unconscious of her own!—
One smiling boy, her last sweet hope, she warms
Hushed on her bosom, circled in her arms,—
Daughter of woe! ere morn, in vain caress'd,
Clung the cold Babe upon thy milkless breast,
With feeble cries thy last sad aid required,
Stretch'd its stiff limbs, and on thy lap expired!—
—Long with wide eye-lids on her Child she gazed,
And long to heaven their tearless orbs she raised;
Then with quick foot and throbbing heart she found
Where Chartreuse open'd deep his holy ground;
Bore her last treasure through the midnight gloom,
And kneeling drop'd it in the mighty tomb;
"I follow next!" the frantic mourner said,
And living plunged amid the festering dead.

Where vast Ontario rolls his brineless tides,
And feeds the trackless forests on his sides,
Fair CASSIA trembling hears the howling woods,

Where Chartreuse. l. 338. During the plague in London, 1665. One pit to receive the dead was dug in the Charter-house, 40 feet long, 16 feet wide, and about 20 feet deep; and in two weeks received 1114 bodies. During this dreadful calamity there were instances of mothers carrying their own children to those public graves, and of people delirious, or in dispair from the loss of their friends, who threw themvelves alive into these pits. Journal of the Plague-year in 1665, printed for E. Nutt. Royal Exchange.

Rolls his brineless tide. l. 343. Some philosophers have believed that the continent of America was not raised out of the great ocean at so early a period of time as the other continents. One reason for this opinion was because the great lakes perhaps nearly as large as the Mediterranean Sea, consist of fresh water. And as the sea-salt seems to have its origin from the destruction of vegetable and animal bodies, washed down by rains and carried by rivers into lakes, or seas. It would seem that this source of sea-salt had not so long existed in that country. There is however a more satisfactory way of explaining this circumstance; which is, that the American lakes lie above the level of the ocean, and are hence perpetually desalited by the rivers which run through them; which is not the case with the Mediterranean, into which a current from the main ocean perpetually passes.

Cassia. l. 345. Ten males, one female. The seeds are black, the stamens gold-colour. This is one of the American fruits, which are annually thrown on the coasts of Norway; and are frequently in so recent a state as to vegetate, when properly taken care of, the fruit of the anacardium, cashew-nut; of cucurbita lagenaria, bottlegourd; of the mimosa scandens,

And trusts her tawny children to the floods.—
Cinctured with gold while *ten* fond brothers stand,
And guard the beauty on her native land,
Soft breathes the gale, the current gently moves,
And bears to Norway's coasts her infant-loves.
—So the sad mother at the noon of night
From bloody Memphis stole her silent flight;
Wrap'd her dear babe beneath her folded vest,
And clasp'd the treasure to her throbbing breast,

---

cocoons; of the piscidia erythrina, logwood-tree, and cocoa-nuts are enumerated by Dr. Tonning. (Amæn. acad. 149) amongst these emigrant seeds. The fact is truly wonderful, and cannot be accounted for but by the existence of under currents in the depths of the ocean; or from vortexes of water passing from one country to another through caverns of the earth.

Sir Hans Sloane has given an account of four kinds of seeds, which are frequently thrown by the sea upon the coasts of the islands of the northern parts of Scotland. Philos. Trans. abridged. V. 3. p. 540. Which seeds are natives of the West Indies, and seem to be brought thither by the gulf-stream, described below. One of these is called by Sir H. Sloane Phaseolus maximus perennis, which is often also thrown on the coast of Kerry in Ireland. Another is called, in Jamaica, Horse-eye-bean; and a third is called Niker in Jamaica. He adds that the Lenticula marina, or Sargosso, grows on the rocks about Jamaica, is carried by the winds and current towards the coast of Florida, and thence into the North-American ocean, where it lies very thick on the surface of the sea.

Thus a rapid current passes from the gulph of Florida to the N.E. along the coast of North-America, known to seamen by the name of the GULF-STREAM. A chart of this was published by Dr. Franklin in 1768, from the information principally of Capt. Folger. This was confirmed by the ingenious experiments of Dr. Blagden published in 1781, who found that the water of the Gulf-stream was from six to eleven degrees warmer than the water of the sea through which it run; which must have been occasioned by its being brought from a hotter climate. He ascribes the origin of this current to the power of the trade-winds, which blowing always in the same direction carry the waters of the atlantic ocean to the westward, till they are stopped by the opposing continent on the west of the Gulf of Mexico, and are thus accumulated there, and run down through the gulf of Florida. Philos. Trans V. 71, p. 335. Governor Pownal has given an elegant map of this Gulf-stream, tracing it from the Gulf of Florida northward as far as Cape Sable in Nova Scotia, and then across the Atlantic ocean to the coast of Africa between the Canary islands and Senegal, increasing in breadth, as it runs, till it occupies five or six degrees of latitude. The Governor likewise ascribes this current to the force of the trade-winds *protruding* the waters westward, till they are opposed by the continent, and accumulated in the Gulf of Mexico. He very ingeniously observes that a great eddy must be produced in the atlantic ocean between this Gulf-stream, and the westerly current protruded by the tropical winds, and in this eddy are found the immense fields of floating vegetables, called Saragosa weeds, and Gulf-weeds, and some light woods, which circulate in these vast eddies, or are occasionally driven out of them by the winds. Hydraulic and nautical observations by Governor Pownal, 1787. Other currents are mentioned by the Governor in this ingenious work, as those in the Indian sea, northward of the line, which are ascribed to the influence of the Monsoons. It is probable that in process of time, the narrow tract of land on the west of the gulf of Mexico may be worn away by this elevation of water dashing against it; by which this immense current would cease to exist; and a wonderful change take place in the Gulf of Mexico, and West Indian islands, by the subsiding of the sea.

With soothing whispers hushed its feeble cry,
Pressed the soft kiss, and breathed the secret sigh.—
—With dauntless step she seeks the winding shore,
Hears unappall'd the glimmering torrents roar;
With Paper-flags a floating cradle weaves,
And hides the smiling boy in Lotus-leaves;
Gives her white bosom to his eager lips,
The salt tears mingling with the milk he sips;
Waits on the reed-crown'd brink with pious guile,
And trusts the scaly monsters of the Nile.—
—Erewhile majestic from his lone abode,
Embassador of Heaven, the Prophet trod;
Wrench'd the red Scourge from proud Opression's hands,
And broke, curst Slavery! thy iron bands.

Hark! heard ye not that piercing cry,
Which shook the waves, and rent the sky!—
E'en now, e'en now, on yonder Western shores
Weeps pale Despair, and writhing Anguish roars:
E'en now in Afric's groves with hideous yell
Fierce SLAVERY stalks, and slips the dogs of hell;
From vale to vale the gathering cries rebound,
And sable nations tremble at the sound!—
—YE BANDS OF SENATORS! whose suffrage sways
Britannia's realms, whom either Ind obeys;
Who right the injured, and reward the brave,
Stretch your strong arm, for ye have power to save!
The close recesses of the heart within,
Stern CONSCIENCE sits, the arbiter of Sin;
With still small voice the plotts of Guilt alarms,
Lights his dark mind, his lifted hand disarms;
But, wrap'd in night with terrors all his own,
He speaks in thunder, when the deed is done.
Hear him, ye Senates! hear this truth sublime,
"He, who allows oppression, shares the crime."

No radiant pearl, which crested Fortune wears,
No gem, that twinkling hangs from Beauty's ears,
Not the bright stars, which night's blue arch adorn,
Nor vernal suns, that gild the rising morn,

Shine with such lustre as the tear, that breaks
For other's woe down Virtue's manly cheeks.

Here ceased the MUSE, and drop'd her tuneful shell,
Tumultuous woes her rising bosom swell,
O'er her flush'd cheek her gauzy veil she throws,
Folds her white arms, and bends her laurel'd brows;
For human guilt awhile the Goddess sighs,
And human sorrows dim celestial eyes.

INTERLUDE. III.

Bookseller. Poetry has been called a sister-art both to Painting, and to Music; 1
I wish to know, what are the particulars of their relationship?

Poet. It has been already observed, that the principal part of the language of 2
poetry consists of those words, which are expressive of the ideas, which we originally receive by the organ of sight; and in this it nearly indeed resembles painting; which can express itself in no other way, but by exciting the ideas or sensations belonging to the sense of vision. But besides this essential similitude in the language of the poetic pen and pencil, these two sisters resemble each other, if I may so say, in many of their habits and manners. The painter to produce a strong effect makes a few parts of his picture large, distinct, and luminous, and keeps the remainder in shadow, or even beneath its natural size and colour, to give eminence to the principal figure. This is similar to the common manner of poetic composition, where the subordinate characters are kept down, to elevate and give consequence to the hero or heroine of the piece.

In the south aile of the cathedral church at Lichfield, there is an antient monu- 3
ment of a recumbent figure; the head and neck of which lie on a roll of matting in a kind of niche or cavern in the wall; and about five feet distant horizontally in another opening or cavern in the wall are seen the feet and ankles, with some folds of garment, lying also on a matt; and tho' the intermediate space is a solid stone-wall, yet the imagination supplies the deficiency, and the whole figure seems to exist before our eyes. Does not this resemble one of the arts both of the painter and the poet? The former often shews a muscular arm amidst a group of figures, or an impassioned face; and, hiding the remainder of the body behind other objects, leaves the imagination to compleat it. The latter describing a single feature or attitude in picturesque words produces before the mind an image of the whole.

I remember seeing a print, in which was represented a shriveled hand stretched 4
through an iron grate, in the stone floor of a prisonyard, to reach at a mess of porrage; which affected me with more horrid ideas of the distress of the prisoner in the dungeon below, than could have been perhaps produced by an exhibition of the whole person. And in the following beautiful scenery from the Midsummer-night's dream, (in which I have taken the liberty to alter the place of a comma,)

the description of the swimming step and prominent belly bring the whole figure before our eyes, with the distinctness of reality.

> When we have laugh'd to see the sails conceive, 5
> And grow big-bellied with the wanton wind; 6
> Which she with pretty and with swimming gate, 7
> Following her womb, (then rich with my young squire) 8
> Would imitate, and sail upon the land. 9

There is a third sister-feature, which belongs both to the pictorial and poetic art; 10
and that is the making sentiments and passions visible, as it were, to the spectator; this is done in both arts by describing or portraying the effects or changes, which those sentiments or passions produce upon the body. At the end of the unaltered play of Lear, there is a beautiful example of poetic painting; the old King is introduced as dying from grief for the loss of Cordelia; at this crisis, Shakespear, conceiving the robe of the King to be held together by a clasp, represents him as only saying to an attendant courtier in a faint voice, "Pray, Sir, undo this button,—thank you, Sir," and dies. Thus by the art of the poet, the oppression at the bosom of the dying King is made visible, not described in words.

B. What are the features, in which these Sister-arts do not resemble each other? 11

P. The ingenious Bishop Berkley in his treatise on Vision, a work of great abil- 12
ity, has evinced; that the colours, which we see, are only a language suggesting to our minds the ideas of solidity and extension; which we had before received by the sense of touch. Thus when we view the trunk of a tree, our eye can only acquaint us with the colours or shades; and from the previous experience of the sense of touch, these suggest to us the cylindrical form, with the prominent or depressed wrinkles on it. From hence it appears, that there is the strictest analogy between colours and sounds; as they are both but languages, which do not represent their correspondent ideas; but only suggest them to the mind from the habits or associations of previous experience. It is therefore reasonable to conclude, that the more artificial arrangements of these two languages by the poet and the painter bear a similar analogy.

But in one circumstance the Pen and the Pencil differ widely from each other; 13
and that is the quantity of Time, which they can include in their respective representations. The former can unravel a long series of events, which may constitute the history of days or years; while the latter can exhibit only the actions of a moment. The Poet is happier in describing successive scenes, the Painter in representing stationary ones. Both have their advantages.

Where the passions are introduced, as the Poet on one hand has the power gradu- 14
ally to prepare the mind of his reader by previous climacteric circumstances; the Painter on the other hand can throw stronger illumination and distinctness on the principal moment or catastrophy of the action; besides the advantage he has in using an universal language, which can be *read* in an instant of time. Thus where a great number of figures are all seen together, supporting or contrasting each other, and

contributing to explain or aggrandize the principal effect, we view a picture with agreeable surprize, and contemplate it with unceasing admiration. In the representation of the sacrifice of Jeptha's Daughter, a print done from a painting of Ant. Coypel, at one glace of the eye we read all the interesting passages of the last act of a well-written tragedy; so much poetry is there condensed into a moment of time.

B. Will you now oblige me with an account of the relationship between Poetry, 15
and her other sister, Music?

P. In the poetry of our language I dont think we are to look for any thing anal- 16
ogous to the notes of the gamut; for except perhaps in a few exclamations, or interrogations, we are at liberty to raise or sink our voice an octave or two at pleasure, without altering the sense of the words. Hence if either poetry or prose be read in melodious tones of voice, as is done in recitativo, or in chaunting, it must depend on the speaker, not on the writer. For though words may be selected, which are less harsh than others; that is, which have fewer sudden stops or abrupt consonants amongst the vowels, or with fewer sibilant letters, yet this does not constitute melody; which consists of agreeable successions of notes referable to the gamut; or harmony, which consists of agreeable combinations of them. If the Chinese language has many words of similar articulation, which yet signify different ideas, when spoken in a higher or lower musical note, as some travelers affirm; it must be capable of much finer effect in respect to the audible part of poetry than any language, we are acquainted with.

There is however another affinity, in which poetry and music more nearly 17
resemble each other than has generally been understood, and that is in their measure or time. There are but two kinds of time acknowledged in modern music, which are called *triple time*, and *common time*. The former of these is divided by bars, each bar containing three crotchets, or a proportional number of their subdivisions into quavers, and semiquavers. This kind of time is analogous to the measure of our heroic or iambic verse. Thus the two following couplets are each of them divided into five bars of *triple time*, each bar consisting of two crotchets and two quavers; nor can they be divided into bars analogous to *common time* without the bars interfering with some of the crotchets, so as to divide them.

3/4 Soft-warbling beaks | in each bright blos | som move, 18
And vo | cal rosebuds thrill | the inchanted grove, | 19

In these lines there is a quaver and a crotchet alternately in every bar; except in 20
the last, in which *the in* make two semiquavers; the *e* is supposed by grammarians to be cut off, which any one's ear will readily determine not to be true.

3/4 Life buds or breathes | from Indus to | the poles, 21
And the | vast surface kind | les, as it rolls.| 22

In these lines there is a quaver and a crotchet alternately in the first bar; a 23
quaver two crotchets and a quaver make the second bar. In the third bar there is a quaver a crotchet and a rest after the crotchet, that is after the word *poles*,

and two quavers begin the next line. The fourth bar consists of quavers and crotchets alternately. In the last bar there is a quaver, and a rest, after it, viz. after the word *kindles*, and then two quavers and a crotchet. You will clearly perceive the truth of this if you prick the musical characters above mentioned under the verses.

The *common time* of musicians is divided into bars, each of which contains 24
four crotchets, or a proportional number of their subdivision into quavers and semiquavers. This kind of musical time is analogous to the dactyle verses of our language, the most popular instances of which are in Mr. Anstie's Bath-guide. In this kind of verse the bar does not begin till after the first or second syllable; and where the verse is quite compleat, and written by a good ear, these first syllables added to the last compleat the bar, exactly in this also corresponding with many pieces of music;

2/4 Yet | if one may guesss by the | size of his calf, Sir, 25
He | weighs about twenty-three | stone and a half, Sir. 26

2/4 Master | Mamozet's head was not | finished so soon, 27
For it | took up the barber a | whole afternoon. 28

In these lines each bar consists of a crotchet, two quavers, another crotchet, and 29
two more quavers: which are equal to four crotchets, and like many bars of *common time* in music may be subdivided into two in beating time without disturbing the measure.

The following verses from Shenstone belong likewise to common time. 30

2/4 A | river or a sea | 31
Was to him a dish | of tea, 32
And a king | dom bread and butter. 33

The first and second bars consist each of a crotchet, a quaver, a crotchet, a quaver, 34
a crotchet. The third bar consists of a quaver, two crotchets, a quaver, a crotchet. The last bar is not compleat without adding the letter A, which begins the first line, and then it consists of a quaver, a crotchet, a quaver, a crotchet, two quavers.

It must be observed, that the crotchets in triple time are in general played by 35
musicians slower than those of common time, and hence minuets are generally pricked in triple time, and country dances generally in common time. So the verses above related, which are analogous to *triple time*, are generally read slower, than those analogous to *common time*; and are thence generally used for graver compositions. I suppose all the different kinds of verses to be found in our odes, which have any measure at all, might be arranged under one or other of these two musical times; allowing a note or two sometimes to preceed the commencement of the bar, and occassional rests, as in musical compositions; if this was attended to by those, who set poetry to music, it is probable, the sound and

sense would oftener coincide. Whether these musical times can be applied to the lyric and heroic verses of the greek and latin poets, I do not pretend to determine; certain it is that the dactyle verse of our language, when it is ended with a double rhime, much resembles the measure of Homer and Virgil, except in the length of the lines.

B. Then there is no relationship between the other two of these sister-ladies, 36
Painting and Music?

P. There is at least a mathematical relationship, or perhaps I ought rather 37
to have said a metaphysical relationship between them. Sir Isaac Newton has observed, that the breadths of the seven primary colours in the Sun's image refracted by a prism are proportional to the seven musical notes of the gamut, or to the intervals of the eight sounds contained in an octave, that is proportional to the following numbers:

Sol.	La.	Fa.	Sol.	La.	Mi.	Fa.	Sol.	38
Red.	Orange.	Yellow.	Green.	Blue.	Indigo.	Violet.		
1/9	1/16	1/10	1/9	1/16	1/16	1/9		

Newton's Optics Book I. part 2. prop. 3 and 6. Dr. Smith in his Harmonics has 39
an explanatory note upon this happy discovery, as he terms it, of Newton. Sect. 4. Art. 7.

From this curious coincidence it has been proposed, to produce a luminous 40
music, consisting of successions or combinations of colours, analogous to a tune in respect to the proportions above mentioned. This might be performed by a strong light, made by means of Mr. Argand's lamps, passing through coloured glasses, and falling on a defined part of a wall; with moveable blinds before them, which might communicate with the keys of a Harpsichord; and thus produce at the same time visible and audible music in unison with each other.

The execution of this idea is said by Mr. Guyot to have been attempted by 41
Father Cassel without much success.

If this should be again attempted, there is another curious coincidence 42
between sounds and colours discovered by Dr. Darwin of Shrewsbury, and explained in a paper on what he calls *Ocular Spectra* in the Philosophical Transactions, Vol. 76, which might much facilitate the execution of it. In this treatise the Doctor has demonstrated, that we see certain colours, not only with greater ease and distinctness, but with relief and pleasure after having for some time contemplated other certain colours; as green after red, or red after green; orange after blue, or blue after orange; yellow after violet, or violet after yellow. This he shews arises from the *ocular spectrum* of the colour last viewed coinciding with the *irritation* of the colour now under contemplation. Now as the pleasure we receive from the sensation of melodious notes, independent of the previoius associations of agreeable ideas with them, must arise from our hearing some proportions of sounds after others more easily,

distinctly, or agreeably; and as there is a coincidence between the proportions of the primary colours, and the primary sounds, if they may be so called. He argues, that the same laws must govern the sensations of both. In this circumstance therefore consists the sisterhood of Music and Painting; and hence they claim a right to borrow metaphors from each other; musicians to speak of the brilliancy of sounds, and the light and shade of a concerto; and Painters of the harmony of colours, and the tone of a picture. Thus it was not quite so absurd, as was imagined, when the blind man asked, if the colour scarlet was like the sound of a trumpet. As the coincidence or opposition of these *ocular spectra*, (or colours which remain in the eye after having for some time contemplated a luminous object;) are more easily and more accurately ascertained, now their laws have been investigated by Dr. Darwin, than the *relicts* of evanescent sounds upon the ear; it is to be wished that some ingenious musician would further cultivate this curious field of science: for if visible music can be agreeably produced, it would be more easy to add sentiment to it by the representations of groves and Cupids and sleeping nymphs amid the changing colours, than is commonly done by words of audible music.

B. You mentioned the greater length of the verses of Homer and Virgil. Had 43
not these poets great advantage in the superiority of their languages compared to our own?

P. It is probable, that the introduction of philosophy into a country must gradu- 44
ally affect the language of it; as philosophy converses in more appropriated and abstracted terms; and thus by degrees eradicates the abundance of metaphor, which is used in the more early ages of society. Otherwise though the greek compound words have more vowels in proportion to their consonants than the english ones, yet the modes of compounding them are less general; as may be seen by variety of instances given in the preface of the Translators, prefixed to the System of Vegetables by the Lichfield society; which happy property of our own language rendered that translation of Linneus as expressive and concise, perhaps more so, than the original.

And in one respect, I believe, the english language serves the purposes of poetry 45
better than the antient ones; I mean in the greater case of producing personifications: For as our nouns have in general no genders affixed to them in prose-compositions, and in the habits of conversation; they become easily personified only by the addition of a masculine or feminine pronoun, as

Pale Melancholy sits, and round *her* throws 46
A deathlike silence, and a dread repose. 47
Pope's Abelard. 48

And secondly as most of our nouns have the article *a* or *the* prefixed to them in 49
prose-writing, and in conversation, they in general become personified even by the omission of these articles; as in the bold figure of Shipwreck in Miss Seward's Elegy on Capt. Cook,

But round the steepy rocks, and dangerous strand 50
Rolls the white surf, and SHIPWRECK guards the land. 51

Add to this, that if the verses in our heroic poetry be shorter than those of 52
the ancients; our words likewise are shorter; and in respect to their measure or time; which has erroneously been called melody and harmony; I doubt from what has been said above, whether we are so much inferior, as is generally believed. Since many passages, which have been stolen from antient Poets, have been translated into our language without loosing any thing of the beauty of the versification.

B. I am glad to hear you acknowledge the thefts of the modern poets from the 53
antient ones, whose works I suppose have been reckoned lawful plunder in all ages. But have not you borrowed epithets, phrases, and even half a line occasionally from modern poems?

P. It may be difficult to mark the exact boundary of what should be termed 54
plagiarism: where the sentiment and expression are both borrowed without due acknowledgement, there can be no doubt;—single words on the contrary taken from other authors cannot convict a writer of Plagiarism, they are lawful game, wild by nature, the property of all who can capture them;—and perhaps a few common flowers of speech may be gathered, as we pass over our neighbour's inclosure, without stigmatising us with the title of thieves; but we must not therefore plunder his cultivated fruit.

The four lines at the end of the plant Upas are imitated from Dr. Yonge's Night- 55
thoughts. The line in the episode adjoined to Cassia "The salt tear mingling with the milk he sips," is from an interesting and humane passage in Langhorn's Justice of Peace. There are probably many others; which, if I could recollect them, should here be acknowledged. As it is,—like exotic plants, their mixture with the native ones, I hope, adds beauty to my Botanic Garden.—And such as it is, *Mr. Bookseller*, I now leave it to you to desire the Ladies and Gentlemen to walk in; but please to apprise them, that, like the spectators at an unskilful exhibition in some village-barn, I hope they will make Good Humour one of their party; and thus themselves supply the defects of the representation.

THE LOVES OF THE PLANTS.
CANTO IV

Now the broad Sun his golden orb unshrouds,
Flames in the west, and paints the parted clouds;
O'er heaven's wide arch refracted lustres flow,
And bend in air the many-colour'd bow.—
—The tuneful Goddess on the glowing sky
Fix'd in mute extacy her glistening eye;
And then her lute to sweeter tones she strung,
And swell'd with softer chords the Paphian song.
Long ailes of Oaks return'd the silver sound,
And amorous Echoes talk'd along the ground;
Pleas'd Lichfield listen'd from her sacred bowers,
Bow'd her tall groves, and shook her stately towers.

Nymph! not for thee the radiant day returns,
Nymph! not for thee the golden solstice burns,
Refulgent Cerea!—at the dusky hour

Pleas'd Lichfield. l. 11. The scenery described at the beginning of the first part, or economy of vegetation, is taken from a botanic garden about a mile from Lichfield.

Cerea. l. 15. Cactus grandiflorus, or Cereus. Twenty males, one female. This flower is native to Jamaica and Veracrux. It expands a most exquisitely beautiful corol, and emits a most fragrant odour for a few hours in the night, and then closes to open no more. The flower is nearly a foot in diameter, the inside of the calyx of a splendid yellow and the numerous petals of a pure white: it begins to open about seven or eight o'clock in the evening, and closes before sun rise in the morning. Martyn's Letters, p. 294. The Cistus labdiniferus, and many other flowers lose their petals after having been a few hours expanded in the day time; for in these plants the stigma is soon impregnated by the numerous anthers; in many flower of the Cistus labdaniferus I observed two or three of the stamens were perpetually bent into contact with the pistil.

The Nyctanthes, called Arabian Jasmine, is another flower, which expands a beautiful corol, and gives out a most delicate perfume during the night, and not in the day in its native country, whence its name; botanical philosophers have not yet explained this wonderful property; perhaps the plant sleeps during the day as some animals do; and its odoriferous glands only emit their fragrance during the expansion of the petals; that is, during its waking hours: the Geranium triste has the same

She seeks with pensive step the mountain-bower,
Bright as the blush of rising morn, and warms
The dull cold eye of Midnight with her charms.
There to the skies she lifts her pencil'd brows,
Opes her fair lips, and breathes her virgin vows;
Eyes the white zenith; counts the suns, that roll
Their distant fires, and blaze around the Pole;
Or marks where Jove directs his glittering car
O'er Heaven's blue vault,—Herself a brighter star.
—There as soft Zephyrs sweep with pausing airs
Thy snowy neck, and part thy shadowy hairs,
Sweet maid of Night! to Cynthia's sober beams
Glows thy warm cheek, thy polish'd bosom gleams.
In crouds around thee gaze the admiring swains,
And guard in silence the enchanted plains.
Drop the still tear, or breathe the impassion'd sigh,
And drink inebriate rapture from thine eye.
Thus, when old Needwood's hoary scenes the Night
Paints with blue shadow, and with milky light;
Where MUNDY pour'd, the listening nymphs among,
Loud to the echoing vales his parting song;
With measured step the Fairy Sovereign treads,
Shakes her high plume, and glitters o'er the meads;
Round each green holly leads her sportive train,
And little footsteps mark the circled plain.
Each haunted rill with silver voices rings,
And Night's sweet bird in livelier accents sings.

Ere the bright star, which leads the morning sky,
Hangs o'er the blushing east his diamond eye,
The chaste TROPÆO leaves her secret bed;

---

property of giving up its fragrance only in the night. The flowers of the Cucurbita lagenaria are said to close when the sun shines upon them. In our climates many flowers, as tragopogon, and hibiscus, close their flowers before the hottest part of the day comes on; and the flowers of some species of cucubalus, and Silene, viscous campion, are closed all day; but when the sun leaves them they expand, and emit a very agreeable scent; whence such plants are termed noctiflora.

Where Mundy. l. 35. Alluding to an unpublished poem by F.N. Mundy, Esq. on his leaving Needwood-Forest.

Tropæolum. l. 45. Majus. Garden Nasturtion, or greater Indian cress. Eight males one female. Miss. E. C. Linneus first observed the tropæolum Majus, to emit sparks or flashes in the mornings before sun-rise, during the months of June or July, and also during the twilight in the evening, but not after total darkness came on; these singular scintillations were shewn to her father and other philosophers, and Mr. Wilcke, a celebrated electrician, believed them to be electric. Lin. Spec. Plantar.

A saint-like glory trembles round her head;
Eight watchful swains along the lawns of night
With amorous steps pursue the virgin-light;
O'er her fair form the electric lustre plays,
And cold she moves amid the lambent blaze.
So shines the glow-fly, when the sun retires,
And gems the night-air with phosphoric fires;
Thus o'er the marsh aerial lights betray,
And charm the unwary wanderer from his way.
So when thy King, Assyria, fierce and proud,
Three human victims to his idol vow'd;
Rear'd a vast pyre before the golden shrine
Of sulphurous coal, and pitch-exsuding pine;—
—Loud roar the flames, the iron nostrils breathe,
And the huge bellows pant and heave beneath;
Bright and more bright the blazing deluge flows,
And white with sevenfold heat the furnace glows.
And now the Monarch fix'd with dread surprize
Deep in the burning vault his dazzled eyes.
"Lo! Three unbound amid the frightful glare,
Unscorched their sandals, and unsinged their hair!
And now a fourth with seraph-beauty bright
Descends, accosts them, and outshines the light!
Fierce flames innocuous, as they step, retire;
And calm they move amid a world of fire!"
He spoke,—to Heaven his arms repentant spread,
And kneeling bow'd his gem-incircled head.

Two Sister-Nymphs, the fair AVENAS, lead
Their fleecy squadrons on the lawns of Tweed;

---

p. 490. Swedish acts for the year 1762. Pulteney's view of Linneus, p. 220. Nor is this more wonderful, than that the electric eel and torpedo should give voluntary shocks of electricity; and in this plant perhaps, as in those animals, it may be a mode of defence, by which it harasses or destroys the night-flying insects, which infest it; and probably, it may emit the same sparks during the day, which must be then invisible. This curious subject deserves further investigation. See Dictamnus. The nectary of this plant grows from what is supposed to be the calyx, and not from the corol.

So shines the glow-fly. l. 51. In Jamaica in some seasons of the year the fire-flies are seen in the evenings in great abundance. When they settle on the ground the bull-frog greedily devours them; which seems to have given origin to a curious though cruel method of destroying these animals; if red-hot pieces of charcoal be thrown towards them in the dusk of the evening, they leap at them, and hastily swallowing them are burnt to death.

Ovena. l. 73. Oat. The numerous families of grasses have all three males, and two females, except Anthoxanthum, which gives the grateful smell to hay, and has but two males. The herbs of this order of vegetables support the countless tribes of graminivorous animals. The feeds of the

Pass with light step his wave-worn banks along,
And wake his Echoes with their silver tongue;
Or touch the reed, as gentle Love inspires,
In notes accordant to their chaste desires.

I.

"Sweet Echo! sleeps thy vocal shell,
Where this high arch o'erhangs the dell;
While Tweed with sun-reflecting streams
Chequers thy rocks with dancing beams?—

II.

"Here may no clamours harsh intrude,
No brawling hound or clarion rude;
Here no fell beast of midnight prowl,
And teach thy tortured cliffs to howl!

III.

"Be thine to pour these vales along
Some artless Shepherd's evening song;
While Night's sweet bird, from yon high spray
Responsive, listens to his lay.

---

smaller kinds of grasses, as of aira, poa, briza, stipa, &c. are the sustenance of many sorts of birds. The seeds of the large grasses, as of wheat, barley, rye, oats, supply food to the human species.

It seems to have required more ingenuity to think of feeding nations of mankind with so small a seed, than with the potatoe of Mexico, or the bread fruit of the southern islands; hence Ceres in Egypt, which was the birth-place of our European arts, was deservedly celebrated amongst their divinities, as well as Osyris, who invented the Plough.

Mr. Wahlborn observes, that as wheat, rye, and many of the grasses, and plaintain, lift up their anthers on long filaments, and thus expose the enclosed fecundating dust to be washed away by the rains, a scarcity of corn is produced by wet summers; hence the necessity of a careful choice of seed-wheat, as that, which had not received the dust of the anthers, will not grow through it may appear well to the eye. The straw of the oat seems to have been the first musical instrument, invented during the pastoral ages of the world, before the discovery of metals. See note on Cistus.

IV.

"And if, like me, some love-lorn maid
Should sing her sorrows to thy shade,
Oh, sooth her breast, ye rocks around!
With softest sympathy of sound."

From ozier bowers the brooding Halcyons peep,
The Swans pursuing cleave the glassy deep,
On hovering wings the wondering Reed-larks play,
And silent Bitterns listen to the lay.—
Three shepherd-swains beneath the beachen shades
Twine rival garlands for the tuneful maids;
On each smooth bark in mystic love-knots frame,
Or on white sands inscribe, the favour'd name.

Warm with sweet blushes bright GALANTHA glows,
And prints with frolic step the melting snows;
O'er silent floods, white hills, and glittering meads
Six rival swains the playful beauty leads,
Chides with her dulcet voice the tardy Spring,
Bids slumbering Zephyr stretch his folded wing,
Wakes the hoarse Cuckoo in his gloomy cave,
And calls the wondering Dormouse from his grave,
Bids the mute Redbreast cheer the budding grove,
And plaintive Ringdove tune her notes to love.

Galanthus. l. 103. Nivalis. Snowdrop. Six males, one female. The first flower that appears after the winter solstice. See Stilingfleet's calendar of Flora.

Some snowdrop-roots taken up in winter, and boiled had the insipid mucilaginous taste of the Orchis, and if cured in the same manner would probably make as good salep. Gmelin in his history of Siberia says the Martigon Lily makes a part of the food of that country, which is of the same natural order as the snowdrop. Some roots of Crocus which I boiled had a disagreeable flavor.

Spring! with thy own sweet smile, and tuneful tongue,
Delighted Bellis calls her infant throng.
Each on his reed astride, the Cherub-train
Watch her kind looks, and circle o'er the plain;
Now with young wonder touch the sliding snail,
Admire his eye-tip'd horns, and painted mail;
Chase with quick step, and eager arms outspread,
The pausing Butterfly from mead to mead;
Or twine green oziers with the fragrant gale,
The azure harebel, and the primrose pale,
Join hand in hand, and in procession gay
Adorn with votive wreaths the shrine of May.
—So moves the Goddess to the Idalian groves,
And leads her gold-hair'd family of Loves.
These, from the flaming furnace, strong and bold
Pour the red steel into the sandy mould;
On tinkling anvils, (with Vulcanian art,)
Turn with hot tongs, and forge the dreadful dart;
The barbed head on whirling jaspers grind,
And dip the point in poison for the mind;
Each polish'd shaft with snow-white plumage wing,
Or strain the bow reluctant to its string.
Those on light wing, above, with busy hands
From bough to bough extend the flowery bands;
Scar the dark beetle, as he wheels on high,
Or catch in silken nets the gilded fly;

The difficulty of raising the Orchis from seed has perhaps been a principal reason of its not being cultivated in this country as an article of food. It is affirmed by one of the Linnean school in the Amænit. academ. that the seeds of Orchis will ripen, if you destroy the new bulb, and that Lily of the Valley, Convallaria, will produce many more seeds and ripen them, if the roots be crouded in a garden pot, so as to prevent them from producing many bulbs. It is probable either of these methods may succeed with these and other bulbous-rooted plants, as snowdrops, and might render their cultivation profitable in this climate. The root of the asphodelus ramosus, branchy asphodel, is used to feed swine in France, and starch is obtained from the alstromeria licta. Memoire's d'Agricult.

Bellis prolifera. l. 114. Hen and chicken Daisy, in this beautiful monster not only the impletion or doubling of the petals takes place, as described in the note on Alcea; but a numerous circlet of less flowers on peduncles, or footstalks, rise from the sides of the calyx, and surround the proliferous parent. The same occurs in Calendula, marigold, in Heracium, hawk-weed, and in Scabiosa, Scabious. Phil. Botan. p. 82.

The fragrant Gale. l. 121. The buds of the Myrica Gale possess an agreeable aromatic fragrance, and might be worth attending to as an article of the materia medica. Mr. Sparman suspects, that the green waxlike substance, with which at certain times of the year the berries of the Myrica cerifera, or candle-berry Myrtle, are covered, are deposited there by insects. It is used by the inhabitants for making candles, which he says burn rather better than those made of tallow. *Voyage to the Cape, V. I. p.* 345.

Call the young Zephyrs to their fragrant bowers,
And stay with kisses sweet the Vernal Hours.

Where, as proud Masson rises rude and bleak,
And with mishapen turrets crests the Peak,
Old Matlock gapes with marble jaws, beneath,
And o'er scar'd Derwent bends his flinty teeth;
Deep in wide caves below the dangerous soil
Blue sulphurs flame, imprison'd waters boil.
Impetuous steams in spiral columns rise
Through rifted rocks, impatient for the skies;
Or o'er bright seas of bubbling lavas blow,
As heave and toss the billowy fires below;
Condensed on high, in wandering rills they glide
From Masson's dome, and burst his sparry side;
Round his grey towers, and down his fringed walls,
From cliff to cliff, the liquid treasure falls;
In beds of stalactite, bright ores among,
O'er corals, shells, and crystals, winds along;
Crusts the green mosses, and the tangled wood,
And sparkling plunges to its parent flood.
—O'er the warm wave a smiling youth presides,
Attunes its murmurs, its meanders guides,

Deep in wide caves. l. 145. The arguments which tend to shew, that the warm springs of this country are produced from steam raised by deep subterraneous fires, and afterwards condensed between the strata of the mountains, appear to me much more conclusive, than the idea of their being warmed by chemical combination near the surface of the earth. for 1st. their heat has kept accurately the same perhaps for many centuries, certainly as long as we have been possessed of good thermometers; which can not be well explained without supposing, that they are first in a boiling state. For as the heat of boiling water is 212, and that of the internal parts of the earth 48, it is easy to understand, that the steam raised from boiling water after being condensed in some mountain, and passing from thence through a certain space of the cold earth, must be cooled always to a given degree, and it is probable the distance from the exit of the spring, to the place where the steam is condensed, might be guessed by the degree of its warmth.

2. In the dry summer of 1780, when all other springs were either dry, or much diminished, those of Buxton and Matlock, (as I was well informed on the spot,) had suffered no diminution; which proves, that the sources of these warm springs are at great depths below the surface of the earth.

3. There are numerous perpendicular fissures in the rocks of Derbyshire, in which the ores of lead, and copper, are found; and which pass to unknown depths; and might thence afford a passage to steam from great subterranean fires.

4. If these waters were heated by the decomposition of pyrites, there would be some chalybeate, taste, or sulphureous smell in them. See note in part I. on the existence of central fires.

(The blooming Fucus,) in her sparry coves
To amorous Echo sings his *secret* loves,
Bathes his fair forehead in the misty stream,
And with sweet breath perfumes the rising steam.
—So, erst, an Angel o'er Bethesda's springs,
Each morn descending, shook his dewy wings;
And as his bright translucent form He laves,
Salubrious powers enrich the troubled waves.

Amphibious Nymph, from Nile's prolific bed
Emerging Trapa lifts her pearly head;

Fucus. l. 161. Clandestine marriage. A species of Fucus, or of Conferva, soon appears in all basons, which contain water. Dr. Priestly found that great quantities of pure dephlogisticated air were given up in water at the points of this vegetable, particularly in the sunshine, and that hence it contributed to preserve the water in reservoirs from becoming putrid. The minute divisions of the leaves of subaquatic plants, as mentioned in the note on Trapa, and of the gills of fish, seem to serve another purpose besides that of increasing their surface, which has not I believe been attended to, and that is to facilitate the separation of the air, which is mechanically mixed, or chemically dissolved in water by their points or edges. This appears on immersing a dry hairy leaf in water fresh from a pump, innumerable globules like quicksilver appear on almost every point, for the extremities of these points attract the particles of water less forcibly, than those particles attract each other; hence the contained air, whose elasticity was but just balanced by the attractive power of the surrounding particles of water to each other, finds at the point of each fibre a place, where the resistance to its expansion is less; and in consequence it there expands, and becomes a bubble of air. It is easy to foresee that the rays of the sunshine by being refracted and in part reflected by the two surfaces of these minute air-bubbles must impart to them much more heat than to the transparent water; and thus facilitate their ascent by further expanding them, that the points of vegetables attract the particles of water less than they attract each other is seen by the spherical form of dew-drops on the points of grass. See note on Vegetable Respiration in part I.

Trapa. l. 170. Four males, one female. The lower leaves of this plant grow under water, and are divided into minute capillary ramifications; while the upper leaves are broad and round, and have air-bladders in their footstalks to support them above the surface of the water. As the aerial leaves of vegetables do the office of lungs, by exposing a large surface of vessels with their contained fluids to the influence of the air; so these aquatic leaves answer a similar purpose, like the gills of fish; and perhaps gain from water or give to it a similar material. As the material thus necessary to life seems to abound more in air than in water, the subaquatic leaves of this plant and of sisymbrium, œnanthe, ranunculus aquatilis, water crowfoot, and some others, are cut into fine divisions to increase the surface; whilst those above water are undivided. So the plants on high mountains have their upper leaves more divided, as pimpinella, petroselinum, and others, because here the air is thinner, and thence a larger surface of contact is required. The stream of water also passes but once along the gills of fish, as it is sooner deprived of its virtue; whereas the air is both received and ejected by the action of the lungs of land-animals. The whale seems to be an exception to the above, as he receives water, and spouts it out again from an organ, which I suppose to be a respiratory one. As spring-water is nearly of the same degree of heat in all climates, the aquatic plants, which grow in rills or fountains, are found equally in the torrid, temperate, and frigid zones, as water-cress, water-parsnip, ranunculus, and many others.

In warmer climates the watery grounds are usefully cultivated, as with rice, and the roots of some aquatic plants are said to have supplied food, as the antient Lotus in Egypt, which some have supposed

Fair glows her virgin cheek and modest breast,
A panoply of scales deforms the rest;
Her quivering fins and panting gills she hides,
But spreads her silver arms upon the tides;
Slow as she sails, her ivory neck she laves,
And shakes her golden tresses o'er the waves.
Charm'd round the Nymph, in circling gambols glide
Four Nereid-forms, or shoot along the tide;
Now all as one they rise with frolic spring,
And beat the wondering air on humid wing;
Now all descending plunge beneath the main,
And lash the foam with undulating train;
Above, below, they wheel, retreat, advance,
In air and ocean weave the mazy dance;
Bow their quick heads, and point their diamond eyes,
And twinkle to the sun with ever-changing dyes.

With net-wove sash and glittering gorget dress'd,
And scarlet robe lapell'd upon her breast,
Stern ARA frowns, the measured march assumes,
Trails her long lance, and nods her shadowy plumes;
While Love's soft beams illume her treacherous eyes,
And Beauty lightens through the thin disguise.
So erst, when HERCULES, untamed by toil,

---

to be the Nymphæa.—In Siberia the roots of the Butomus, or flowering rush, are eaten, which is well worth further inquiry, as they grow spontaneously in our ditches, and rivers; which at present produce no esculent vegetables; and might thence become an article of useful cultivation. Herodotus affirms that the Egyptian Lotos grows in the Nile, and resembles a Lily. That the natives dry it in the sun, and take the pulp out of it, which grows like the head of a poppy, and bake it for bread. Enterpe. Many grit stones and coals, which I have seen, seem to bear an impression of the roots of the Nymphæa, which are often three or four inches thick, especially the white flowered one.

Arum. 1.189. Cuckow-pint, of the class Gynandria, or masculine ladies. The pistil or female part of the flower rises like a club, is covered above or clothed as it were by the anthers or males; and some of the species have a large scarlet blotch in the middle of every leaf.

The singular and wonderful structure of this flower has occasioned many disputes amongst botanists. See Tourniff. Malpig. Dillen. Rivin. &c. The receptacle is enlarged into a naked club, with the germs at its base, the stamens are affixed to the receptacle amidst the germs, (a natural prodigy) and thus do not need the assistance of elevating filaments. Hence the flower may be said to be inverted. *Families of Plants* translated from Linneus, p 618.

The spadix of this plant is frequently quite white, or coloured, and the leaves liable to be streaked with white, and to have black, or scarlet blotches on them. As the plant has no corol or blossom, it is probable the coloured juices in these parts of the sheath or leaves may serve the same purpose, as the coloured juices in the petals of other flowers; from which I suppose the honey to be prepared. See note on Helleborus. I am informed that those tulip roots, which have a red cuticle, produce red flowers. See Rubia.

Own'd the soft power of DEJANIRA's smile;
His lion-spoils the laughing Fair demands,
And gives the distaff to his awkward hands;
O'er her white neck the bristly mane she throws,
And binds the gaping whiskers on her brows;
Fits round her slender waste the shaggy vest,
And clasps the velvet paws across her breast.
Next with soft hands the knotted club she rears,
Heaves up from earth, and on her shoulder bears.
Onward with loftier step the Beauty treads,
And trails the brinded ermine o'er the meads;
Wolves bears and pards forsake the affrighted groves,
And grinning Satyrs tremble, as she moves.

CARYO's sweet smile DIANTHUS proud admires,
And gazing burns with unallow'd desires;
With sighs and sorrows her compassion moves,
And wins the damsel to illicit loves.
The Monster-offspring heirs the father's pride,

Dianthus. l. 207. Superbus. Proud Pink. There is a kind of pink called Fairchild's mule, which is here supposed to be produced between a Dianthus superbus, and the Caryophyllus, Clove. The Dianthus superbus emits a most fragrant odour, particularly at night. Vegetable mules supply an irrefragable argument in favour of the sexual system of botany. They are said to be numerous, and like the mules of the animal kingdom not always to continue their species by seed. There is an account of a curious mule from the Antirrhinum linaria, Toad-flax, in the Amænit. academ. V. I. No. 3. and many hybrid plants described in No. 32. The Urtica alienata is an evergreen plant, which appears to be a nettle from the male flowers, and a Pellitory (Parietaria) from the female ones and the fruit; and is hence between both. Murray. Syst. Veg. Amongst the English indigenous plants the veronica hybrida mule Speedwel is supposed to have originated from the officinal one, and the spiked one, and the Sibthorpia europæa to have for its parents the golden saxifrage, and marsh pennywort. Pulteney's view of Linneus, p. 250. Mr. Graberg, Mr. Schreber, and Mr. Ramstrom seem of opinion, that the internal structure or parts of fructification in mule plants resemble the female parent, but that the habit or external structure resembles the male parent. See treatises under the above names in V. 6. Amænit. academic. The mule produced from a horse and the ass resembles the horse externally with his ears, main, and tail; but with the nature or manners of an ass: but the Hinnus, or creature produced from a male ass, and a mare, resembles the father externally, in stature, ash-colour, and the black cross, but with the nature or manners of a horse. The breed from Spanish rams and Swedish ewes resembled the Spanish sheep in wool, stature and external form; but was as hardy as the Swedish sheep; and the contrary of those, which were produced from Swedish rams and Spanish ewes. The offspring from the male goat of Angora and the Swedish female goat had long soft camel's hair; but that from the male Swedish goat, and the female one of Angora, had no improvement of their wool. An English ram without horns, and a Swedish horned ewe produced sheep without horns, Amæn. academ. V. 6. p. 13.

Mask'd in the damask beauties of the bride.
So, when the Nightingale in eastern bowers
On quivering pinion woos the Queen of flowers;
Inhales her fragrance, as he hangs in air,
And melts with melody the blushing fair;
Half-rose, half-bird, a beauteous Monster springs,
Waves his thin leaves, or claps his glossy wings;
Long horrent thorns his mossy legs surround,
And tendril-talons root him to the ground;
Green films of rind his wrinkled neck o'erspread,
And crimson petals crest his curled head;
Soft warbling beaks in each bright blossom move,
And vocal Rosebuds thrill the enchanted grove!—
Admiring Evening stays her beamy star,
And still Night listens from his ebon car;
While on white wings descending Houries throng,
And drink the floods of odour and of song.

When from his golden urn the Solstice pours
O'er Afric's sable sons the sultry hours;
When not a gale flits o'er her tawny hills,
Save where the dry Harmattan breathes and kills;
When stretch'd in dust her gasping panthers lie,

The dry Harmattan. l. 230. The Harmattan is a singular wind blowing from the interior parts of Africa to the atlantic ocean, sometimes for a few hours, sometimes for several days without regular periods. It is always attended with a fog or haze, so dense as to render those objects invisible, which are at the distance of a quarter of a mile, the sun appears through it only about noon, and then of a dilute red, and very minute particles subside from the misty air so as to make the grass, and the skins of Negroes appear whiteish. The extreme dryness, which attends this wind or fog without dews, withers and quite dries the leaves of vegetables, and is said of Dr. Lind at some seasons to be fatal and malignant to mankind, probably after much preceeding wet; when it may become loaded with the exhalations from putrid marshes; at other seasons it is said to check epidemic diseases, to cure fluxes and to heal ulcers, and cutaneous eruptions; which is probably effected by its yielding no moisture to the mouths of the external absorbent vessels, by which the action of the other branches of the absorbent system is increased to supply the deficiency. *Account of the Harmattan. Phil. Transact. V.* 71.

In many circumstance this wind seems much to resemble the dry fog, which covered most parts of Europe for many weeks in the summer of 1780, which has been supposed to have had a volcanic origin, as it succeeded the violent eruption of Mount Hecla, and its neighbourhood. From the subsidence of a white powder it seems probable, that the Harmattan has a similar origin, from the unexplored mountains of Africa. Nor is it improbable, that the epidemic coughs, which occasionally traverse immense tracts of country, may be the products of volcanic eruptions; nor impossible, that at some future time contagious miasmata may be thus emitted from subterraneous furnaces, in such abundance as to contaminate the whole atmposphere, and depopulate the earth!

And writh'd in foamy folds her serpents die;
Indignant Atlas mourns his leafless woods,
And Gambia trembles for his sinking floods;
Contagion stalks along the briny sand,
And Ocean rolls his sickening shoals to land.
—Fair CHUNDA smiles amid the burning waste,
Her brow unturban'd, and her zone unbrac'd;
Ten brother-youths with light umbrella's shade,
Or fan with busy hands the panting maid;
Loose wave her locks, disclosing, as they break,
The rising bosom and averted cheek;
Clasp'd round her ivory neck with studs of gold
Flows her thin vest in many a silky fold;
O'er her light limbs the dim transparence plays,
And the fair form, it seems to hide, betrays.

Where leads the northern Star his lucid train
High o'er the snow-clad earth, and icy main,
With milky light the white horizon streams,
And to the moon each sparkling mountain gleams.—
Slow o'er the printed snows with silent walk
Huge shaggy forms across the twilight stalk;
And ever and anon with hideous sound

His sickening shoals. l. 236. Mr. Marsden relates, that in the island of Sumatra, during the November of 1775, the dry monsoons, or S.E. winds, continued so much longer than usual, that the large rivers became dry; and prodigious quantities of sea-fish, dead and dying, were seen floating for leagues on the sea, and driven on the beach by the tides. This was supposed to have been caused by the great evaporation, and the deficiency of fresh water rivers having rendered the sea too salt for its inhabitants. The season then became so sickly as to destroy great numbers of people both foreigners and natives. Phil. Trans. V. 71. p. 384.

Chunda. l. 237. Chundali Borrum is the name, which the natives give to this plant; it is the Hedysarum movens, or moving plant; its class is two brotherhoods ten males. Its leaves are continually in spontaneous motion, some rising and others falling, and others whirling circularly by twisting their stems; this spontaneous movement of the leaves, when the air is quite still, and very warm, seems to be necessary to the plant, as perpetual respiration is to animal life.

There are many other instances of spontaneous movements of the parts of vegetables. In the Marchantia polymorpha some yellow wool proceeds from the flower-bearing anthers, which moves spontaneously in the anther, while it drops its dust like atoms. Murray. Syst. Veg. See note on Collinsonia for other instances of vegetable spontaneity. Add to this, that as the sleep of animals consists in a suspension of voluntary motion, and as vegetables are likewise subject to sleep; there is reason to conclude, that the various actions of opening and closing their petals and foliage may be justly ascribed to a voluntary power: for without the faculty of volition, sleep would not have been necessary to them.

Burst the thick ribs of ice, and thunder round.—
There, as old Winter flaps his hoary wing,
And lingering leaves his empire to the Spring,
Pierced with quick shafts of silver-shooting light
Fly in dark troops the dazzled imps of night.—
"Awake my love!" enamour'd MUSCHUS cries,
"Stretch thy fair limbs, refulgent Maid! arise;
Ope thy sweet eye-lids to the rising ray,
And hail with ruby lips returning day.
Down the white hills dissolving torrents pour,
Green springs the turf, and purple blows the flower;
His torpid wing the Rail exulting tries,
Mounts the soft gale, and wantons in the skies;
Rise, let us mark how bloom the awaken'd groves,
And 'mid the banks of roses *hide* our loves."

Night's tinsel beams on smooth Lock-lomond dance,
Impatient ÆGA views the bright expanse;—
In vain her eyes the passing floods explore,
Wave after wave rolls freightless to the shore.
—Now dim amid the distant foam she spies

Burst of thick ribs of ice. l. 254. The violent cracks of ice heard from the Glaciers seem to be caused by some of the snow being melted in the middle of the day; and the water thus produced running down into vallies of ice, and congealing again in a few hours forces off, by its expansion, large precipices from the ice-mountains.

Muschus. l. 259. Corallinus, or lichen rangiferinus. Coral-moss. Clandestine marriage. This moss vegetates beneath the snow, where the degree of heat is always about 40, that is in the middle between the freezing point, and the common heat of the earth; and is for many months of the winter the sole food of the rain-deer; who digs furrows in the snow to find it; and as the milk and flesh of this animal is almost the only sustenance, which can be procured during the long winters of the higher latitudes, this moss may be said to support some millions of mankind.

The quick vegetation, that occurs on the solution of the snows in high latitudes, appears very astonishing; it seems to arise from two causes, 1. the long continuance of the approaching sun above the horizon, 2. the increased irritability of plants, which have been long exposed to the cold. See notes on part I.

All the water-fowl on the lakes of Siberia are said by professor Gmelin to retreat southwards on the commencement of the frosts except the Rail; which sleeps buried in the snow. Account of Siberia.

Æga. l. 270. Conferva ægagropila. It is found loose in many lakes in a globular form, from the size of a walnut to that of a melon, much resembling the balls of hair found in the stomachs of cows; it adheres to nothing, but rolls from one part of the lake to another. The Conferva vagabunda dwells on the European seas, travelling along in the midst of the waves; (Spec. Plant.) these may not improperly be called itinerant vegetables. In a similar manner the Fucus natans (swimming) strikes no roots into the earth, but floats on the sea, in very extensive masses, and may be said to be a plant of passage, as it is wafted by the winds from one shore to another.

A rising speck,—"'tis he! 'tis he!" she cries;
As with firm arms he beats the streams aside,
And cleaves with rising chest the tossing tide,
With bended knee she prints the humid sands,
Up-turns her glistening eyes, and spreads her hands;
—"'Tis he, 'tis he!—My Lord, my life, my love!—
Slumber ye winds; ye billows, cease to move!
Beneath his arms your buoyant plumage spread,
Ye Swans! Ye Halcyons, hover round his head!"—
—With eager step the boiling surf she braves,
And meets her refluent lover in the waves;
Loose o'er the flood her azure mantle swims,
And the clear stream betrays her snowy limbs.

So on her sea-girt tower fair HERO stood
At parting day, and mark'd the dashing flood;
While high in air, the glimmering rocks above,
Shone the bright lamp, the pilot-star of Love.
—With robe outspread the wavering flame behind
She kneels, and guards it from the shifting wind;
Breathes to her Goddess all her vows, and guides
Her bold LEANDER o'er the dusky tides;
Wrings his wet hair, his briny bosom warms,
And clasps her panting lover in her arms.

Deep, in wide caverns and unfathom'd cells,
Daughter of Earth, the chaste TRUFFELIA dwells;
On silvery beds, of soft asbestus wove,
Meets her Gnome-husband, and avows her love.
—*High* o'er her couch impending diamonds blaze,
And branching gold the crystal roof inlays;
With verdant light the modest emeralds glow,
Blue sapphires glare, and rubies blush, *below;*
Light piers of lazuli the dome surround,
And pictured mochoes tesselate the ground;
O'er the gay ailes delighted Cupids stray,

Truffelia. l. 298. (Lycoperdon Tuber) Truffle. Clandestine marriage. This fungus never appears above ground, requiring little air and perhaps no light. It is found by dogs or swine, who hunt it by the smell. Other plants, which have no buds or branches on their stems, as the grasses, shoot out numerous stoles or scions underground; and this the more, as their tops or herbs are eaten by cattle; and thus preserve themselves.

And shed from odorous lamps celestial day.—

So, cavern'd round in vast Polandish mines,
With crystal walls a gorgeous city shines;
Scoop'd in the briny rock long streets extend
Their hoary course, and glittering domes ascend.
Down the bright steeps, emerging into day,
Impetuous fountains burst their headlong way,
O'er milk-white vales in ivory channels spread,
And wondering seek their subterranean bed.
Form'd in pellucid salt with chissel nice,
The pale lamp glimmering through the sculptur'd ice,
With wild reverted eyes fair LOTTA stands,
And spreads to heaven, in vain, her glassy hands;
Cold dews condense upon her pearly breast,
And the big tear rolls lucid down her vest.
Far-gleaming o'er the town, transparent fanes
Rear their white towers, and wave their golden vanes;
Long lines of Lustres pour their trembling rays,
And the bright vault returns the mingled blaze.

Closed in an azure fig by fairy spells,
Bosom'd in down, fair CAPRI-FICA dwells;—

Scoop'd in the briny rock. l. 311. There is a town in the immense salt-mines of Crackow in Poland, with a market-place, a river, a church, and a famous statue, (here supposed to be Lot's wife) by the moist or dry appearance of which the subterranean inhabitants know, when the weather is fair above ground, these immense masses of rock salt seem to have been produced by the evaporation of sea-water in the early periods of the world by subterranean fires. Dr. Hutton's Theory of the Earth.

Salts of various kinds are produced from the recrements of animals and vegetable bodies, such as phosphoric, ammoniacal, marine salt, and others; these are washed from the earth by rains, and carried down our rivers into the sea; they seem all here to decompose each other except the marine salt; which has therefore from the beginning of the habitable world been perpetually accumulating; as an article of diet it seems to be simply a stimulus, not containing any nourishment; and like all other unnatural stimuli is not necessary to people in health, and contributes to weaken our system; though it may be useful as a medicine. It seems to be an immediate cause of the sea-scurvy, as these patients get well by the use of fresh provisions; and is probably a remote cause of the scrophula, (which consists in the want of irritability of the absorbent system,) and is therefore useful to these patients; as wine is necessary to those whose stomachs have been weakened by its use. The universality of the use of salt with our food and in our cookery has rendered it difficult to prove the truth of these observations.

Caprificus. l. 328. Wild fig, The fruit of the fig is not a seed-vessel, but a receptacle inclosing the flower within it. As these trees bear some male and others female flowers,

So sleeps in silence the Curculio, shut
In the dark chambers of the cavern'd nut,
Erodes with ivory beak the vaulted shell,
And quits on filmy wings its narrow cell.
So the pleased Linnet in the moss-wove nest,
Waked into life beneath its parent's breast,
Chirps in the gaping shell, bursts forth erelong,
Shakes its new plumes, and tries its tender song.—
—And now the talisman she strikes, that charms
Her husband-Sylph,—and calls him to her arms.—
Quick, the light Gnat her airy Lord bestrides,
With cobweb reins the flying courser guides,
From crystal steeps of viewless ether springs,
Cleaves the soft air on still expanded wings;
Darts like a sunbeam o'er the boundless wave,
And seeks the beauty in her *secret* cave.
So with quick impulse through all nature's frame
Shoots the electric air its subtle flame.

---

immured on all sides by the fruit, the manner of their fecundation was very unintelligible, till Tournefort and Pontedera discovered, that a kind of gnat produced in the male figs, carried the fecundating dust on its wings, (Cynips Psenes Syst. Nat. 919) and penetrating the female fig, thus impregnated the flowers; for the evidence of this wonderful fact see the word Caprification, in Milne's botanical dictionary. The figs of this country are all female, and their seeds not prolific; and therefore they can only be propagated by layers, and suckers.

Monsieur de la Hire has shewn in the Memoir. de l'Academ des Science, that the summer figs at Paris, in Provence, Italy, and Malta have all perfect stamina, and ripen not only their fruits, but their seed; from which seed other fig-trees are raised; but that the stamina of the autumnal figs are abortive, perhaps owing to the want of due warmth. Mr. Milne in his botanical dictionary, (art. Caprification) says that the cultivated fig-trees have a few male flowers placed above the female, within the same covering or receptacle; which in warmer climates perform their proper office, but in colder ones become abortive. And Linneus observes, that some figs have the navel of the receptacle open, which was one reason, that induced him to remove this plant from the class, clandestine marriage, to the class Polygamy. Lin. Spec. Plant.

From all these circumstances I should conjecture, that those female fig-flowers, which are closed on all sides in the fruit or receptacle without any male ones, are monsters; which have been propagated for their fruit, like barberries, and grapes without seeds in them, and that the caprification is either an antient process of imaginary use, and blindly followed in some countries, or that it may contribute to ripen the fig by decreasing its vigour; like cutting off a circle of the bark from the branch of a pear-tree. Tournefort seems inclined to this opinion, who says, that the figs in Provence and at Paris ripen sooner, if their buds be pricked with a straw diped in olive-oil. Plumbs and pears punctured by some insects ripen sooner, and the part round the puncture is sweeter. Is not the honey-dew produced by the puncture of insects? will not wounding the branch of a pear-tree, which is too vigorous, prevent the blossoms from falling off; as from some fig-trees the fruit is said to fall off unless they are wounded by caprification? I had last spring six young trees of the Ischia fig with fruit on them in pots in a stove; on removing them into larger boxes, they protruded very vigorous shoots, and the figs all fell off; which I ascribed to the increased vigour of the plants.

So turns the impatient needle to the pole,
Tho' mountains rise between, and oceans roll.

Where round the Orcades white torrents roar,
Scooping with ceaseless rage the incumbent shore,
Wide o'er the deep a dusky cavern bends
Its marble arms, and high in air impends;
Basaltic piers the ponderous roof sustain,
And steep their massy sandals in the main;
Round the dim walls, and through the whispering ailes
Hoarse breathes the wind, the glittering water boils.
Here the charm'd BYSSUS with his blooming bride
Spreads his green sails, and braves the foaming tide;
The star of Venus gilds the twilight wave,
And lights her votaries to the *secret* cave;
Light Cupids flutter round the nuptial bed,
And each coy sea-maid hides her blushing head.

Where cool'd by rills, and curtain'd round by woods,
Slopes the green dell to meet the briny floods,
The sparkling noon-beams trembling on the tide,
The PROTEUS-LOVER woos his playful bride.

Basaltic piers. l. 353. This description alludes to cave of Fingal in the island of Staffa. The basaltic columns, which compose the Giants causeway on the coast of Ireland, as well as those which support the cave of Fingal are evidently of volcanic origin, as is well illustrated in an ingenious paper of Mr. Keir in the Philos. Trans. Vol. who observed in the glass, which had been long in a fusing heat at the bottom of the pots in the glass houses at Stourbridge, that chrystals were produced of a form similar to the parts of the basaltic columns of the Giant's causeway.

Byssus. l. 357. Clandestine marriage. It floats on the sea in the day, and sinks a little during the night, it is found in caverns on the northern shores, of a pale green colour, and as thin as paper.

The Proteus-lover. l. 366. Conferva-polymorpha. This vegetable is put amongst the cryptogamia, or clandestine marriages, by Linneus; but according to Mr. Ellis the males and females are on different plants. Philos. Trans. V. 57. It twice changes its colour, from red to brown, and then to black; and changes its form by losing its lower leaves, and elongating some of the upper ones, so as to be mistaken by the unskilful for different plants, it grows on the shores of this country.

There is another plant, Medicago polymorpha, which may be said to assume a great variety of shapes; as the seed-vessels resemble sometimes snail-horns, at other times caterpillars with or without long hair upon them, by which means it is probable they sometimes elude the depredations of those insects. Salicornia also assumes an animal similitude. Phil. Bot. p. 87. See note on Rubia.

To win the fair he tries a thousand forms,
Basks on the sands, or gambols in the storms.—
A Dolphin now, his scaly sides he laves,
And bears the sportive damsel on the waves;
She strikes the cymbal, as he moves along,
And wondering Ocean listens to the song.
—And now a spotted pard the Lover stalks,
Plays round her steps, and guards her favour'd walks;
As with white teeth he prints her hand, caress'd,
And lays his velvet paw upon her breast,
O'er his round face her snowy fingers strain
The silken knots, and fit the ribbon-rein.
—And now a Swan, he spreads his plumy sails,
And proudly glides before the fanning gales;
Pleas'd on the flowery brink with graceful hand
She waves her floating lover to the land;
Bright shines his sinuous neck, with crimson beak
He prints fond kisses on her glowing cheek,
Spreads his broad wings, elates his ebon crest,
And clasps the beauty to his downy breast.

A *hundred* virgins join a *hundred* swains,
And fond ADONIS leads the sprightly trains;
Pair after pair, along his sacred groves
To Hymen's fane the bright procession moves;
Each smiling youth a myrtle garland shades,
And wreaths of roses veil the blushing maids;
Light Joys on twinkling feet attend the throng,
Weave the gay dance, or raise the frolic song;
—Thick, as they pass, exulting Cupids fling

Adonis. l. 388. Many males and many females live together in the same flower. It may seem a solecism in language, to call a flower, which contains many of both sexes, an individual; and the more so to call a tree or shrub an individual, which consists of so many flowers. Every tree indeed ought to be considered as a family or swarm of its respective buds; but the buds themselves seem to be individual plants; because each has leaves or lungs appropriated to it; and the bark of the tree is only a congeries of the roots of all these individual buds. Thus hollow oak trees and willows are often seen with the whole wood decayed, and gone; and yet the few remaining branches flourish with vigour; but in respect to the male and female parts of a flower, they do not destroy its individuality any more than the number of paps of a sow, or the number of her cotyleadons, each of which includes one of her young.

The society called the Areoi in the island of Otaheite consists of about 100 males and 100 females who form one promiscuous marriage.

Promiscuous arrows from the sounding string;
On wings of gossamer soft Whispers fly,
And the sly Glance steals side-long from the eye.
—As round his shrine the gaudy circles bow,
And seal with muttering lips the faithless vow,
Licentious Hymen joins their mingled hands,
And loosely twines the meretricious bands.—
Thus where pleased VENUS, in the southern main,
Sheds all her smiles on Otaheite's plain,
Wide o'er the isle her silken net she draws,
And the Loves laugh at all, but nature's laws."—

Here ceased the Goddess,—o'er the silent strings
Applauding Zephyrs swept their fluttering wings;
Enraptured Sylphs arose in murmuring crouds
To air-wove canopies and pillowy clouds;
Each Gnome reluctant sought his earthy cell,
And each bright Floret clos'd her velvet bell.
Then, on soft tiptoe, NIGHT approaching near
Hung o'er the tuneless lyre his sable ear;
Gem'd with bright stars the still ethereal plain,
And bad his Nightingales repeat the strain.

ADDITIONAL NOTES.

Description of the Poison-Tree *in the island of* Java. *Translated from the original Dutch of* N. P. Foersch.

This destructive tree is called in the Malayan language *Bohon-Upas*, and has 1
been described by naturalists. But their accounts have been so tinctured with the *marvellous*, that the whole narration has been supposed to be an ingenious fiction by the generality of readers. Nor is this in the least degree surprising, when the circumstances which we shall faithfully relate in this description are considered.

I must acknowledge, that I long doubted the existence of this tree, until a 2
stricter enquiry convinced me of my error. I shall now only relate simple unadorned facts, of which I have been an eye witness. My readers may depend upon the fidelity of this account. In the year 1774 I was stationed at Batavia, as a surgeon in the service of the Dutch East-India Company. During my residence there I received several different accounts of the Bohon-Upas, and the violent effects of its poison. They all then seemed incredible to me, but raised my curiosity in so high a degree, that I resolved to investigate this subject thoroughly, and to trust only to *my own observations*. In consequence of this resolution, I applyed to the Governor-General, Mr. Petrus Albertus van der Parra, for a pass to travel through the country. My request was granted, and having procured every information, I set out on my expedition. I had procured a recommendation from an old Malayan priest to another priest, who lives on the nearest inhabitable spot to the tree, which is about fifteen or sixteen miles distant. The letter proved of great service to me in my undertaking, as that priest is appointed by the Emperor to reside there, in order to prepare for eternity the souls of those who for different crimes are sentenced to approach the tree and to procure the poison.

The *Bohon-Upas* is situated in the island of *Java*, about twenty-seven leagues 3
from *Batavia*, fourteen from *Soura Charta*, the seat of the Emperor, and between eighteen and twenty leagues from *Tinkjoe*, the present residence of the Sultan of Java. It is surrounded on all sides by a circle of high hills and mountains, and the country round it, to the distance of ten or twelve miles from the tree, is entirely

barren. Not a tree, nor a shrub, nor even the least plant or grass is to be seen. I have made the tour all around this dangerous spot, at about eighteen miles distant from the centre, and I found the aspect of the country on all sides equally dreary. The easiest ascent of the hills is from that part where the old ecclesiastick dwells. From his house the criminals are sent for the poison, into which the points of all warlike instruments are dipped. It is of high value, and produces a considerable revenue to the Emperor.

Account of the manner in which the Poison is procured.

The poison which is procured from the tree is a gum that issues out between 4 the bark and the tree itself, like the *camphor*. Malefactors, who for their crimes are sentenced to die, are the only persons who fetch the poison; and this is the only chance they have of saving their lives. After sentence is pronounced upon them by the judge, they are asked in court, whether they will die by the hands of the executioner, or whether they will go to the Upas tree for a box of poison? They commonly prefer the latter proposal, as there is not only some chance of preserving their lives, but also a certainty, in case of their safe return, that a provision will be made for them in future by the Emperor. They are also permitted to ask a favour from the Emperor, which is generally of a trifling nature, and commonly granted. They are then provided with a silver or tortoiseshell box, in which they are to put the poisonous gum, and are properly instructed how to proceed while they are upon their dangerous expedition. Among other particulars, they are always told to attend to the direction of the winds; as they are to go towards the tree before the wind, so that the effluvia from the tree are always blown from them. They are told, likewise, to travel with the utmost dispatch, as that is the only method of insuring a safe return. They are afterwards sent to the house of the old priest, to which place they are commonly attended by their friends and relations. Here they generally remain some days, in expectation of a favourable breeze. During that time, the ecclesiastic prepares them for their future state by prayers and admonitions.

When the hour of their departure arrives, the priest puts them on a long leather 5 cap, with two glasses before their eyes, which comes down as far as their breast, and also provides them with a pair of leather gloves. They are then conducted by the priest, and their friends and relations, about two miles on their journey. Here the priest repeats his instructions, and tells them where they are to look for the tree. He shows them a hill, which they are told to ascend, and that on the other side they will find a rivulet, which they are to follow, and which will conduct them directly to the Upas. They now take leave of each other, and, amidst prayers for their success the delinquents hasten away.

The worthy old ecclesiastic has assured me, that during his residence there, for 6 upwards of thirty years, he had dismissed above seven hundred criminals in the manner which I have described; and that scarcely two out of twenty have returned.

He showed me a catalogue of all the unhappy sufferers, with the date of their departure from his house annexed, and a list of the offences for which they had been condemned. To which was added a list of those who had returned in safety. I afterwards saw another list of these culprits, at the jail-keeper's at *Soura-Charta*, and found that they perfectly corresponded with each other, and with the different informations which I afterwards obtained.

I was present at some of these melancholy ceremonies, and desired different 7
delinquents to bring with them some pieces of the wood, or a small branch, or some leaves of this wonderful tree. I have also given them silk cords, desiring them to measure its thickness. I never could procure more than two dry leaves that were picked up by one of them on his return; and all I could learn from him concerning the tree itself was, that it stood on the border of a rivulet, as described by the old priest, that it was of a middling size, that five or six young trees of the same kind stood close by it; but that no other shrub or plant could be seen near it; and that the ground was of a brownish sand, full of stones, almost impracticable for travelling, and covered with dead bodies. After many conversations with the old Malayan priest, I questioned him about the first discovery, and asked his opinion of this dangerous tree, upon which he gave me the following answer:

"We are told in our new Alcoran, that, above an hundred years ago, the country 8
around the tree was inhabited by a people strongly addicted to the sins of Sodom and Gomorrha. When the great prophet Mahomet determined not to suffer them to lead such detestable lives any longer, he applyed to God to punish them; upon which God caused this tree to grow out the earth, which destroyed them all, and rendered the country for ever uninhabitable."

Such was the Malayan opinion, I shall not attempt a comment, but must observe, 9
that all the Malayans consider this tree as an holy instrument of the great prophet to punish the sins of mankind, and, therefore, to die of the poison of the Upas is generally considered among them as an honourable death. For that reason I also observed that the delinquents, who were going to the tree, were generally dressed in their best apparel.

This, however, is certain, though it may appear incredible, that from fifteen 10
to eighteen miles round this tree, not only no human creature can exist, but that, in that space of ground, no living animal of any kind has ever been discovered. I have also been assured by several persons of veracity, that there are no fish in the water, nor has any rat, mouse, or any other vermin been seen there; and when any birds fly so near this tree that the effluvia reaches them, they fall a sacrifice to the effects of the poison. This circumstance has been ascertained by different delinquents, who, in their return, have seen the birds drop down, and have picked them up *dead*, and brought them to the old ecclesiastick.

I will here mention an instance which proves them a fact beyond all doubt, and 11
which happened during my stay at Java.

In the year 1775 a rebellion broke out among the subjects of the Massay, a 12
Sovereign prince, whose dignity is nearly equal to that of the Emperor. They

refused to pay a duty imposed upon them by their Sovereign, whom they openly opposed. The Massay sent a body of a thousand troops to disperse the rebels, and to drive them, with their families, out of his dominions. Thus four hundred families, consisting of above sixteen hundred souls, were obliged to leave their native country. Neither the Emperor nor the Sultan would give them protection, not only because they were rebels, but also through fear of displeasing their neighbour, the Massay. In this distressful situation, they had no other resource than to repair to the uncultivated parts round the Upas, and requested permission of the Emperor to settle there. Their request was granted, on condition of their fixing their abode not more than twelve or fourteen miles from the tree, in order not to deprive the inhabitants already settled there at a greater distance of their cultivated lands. With this they were obliged to comply; but the consequence was, that in less than two months their number was reduced to about three hundred. The chiefs of those who remained returned to the Massay, informed him of their losses, and intreated his pardon, which induced him to receive them again as subjects, thinking them sufficiently punished for their misconduct. I have seen and conversed with several of those who survived soon after their return. They all had the appearance of persons tainted with an infectious disorder; they looked pale and weak, and from the account which they gave of the loss of their comrades, of the symptoms and circumstances which attended their dissolution, such as convulsions, and other signs of a violent death, I was fully convinced that they fell victims to the poison.

This violent effect of the poison at so great a distance from the tree, certainly 13
appears surprising, and almost incredible; and especially when we consider that it is possible for delinquents who approach the tree to return alive. My wonder, however, in a great measure, ceased, after I had made the following observations:

I have said before, that malefactors are instructed to go to the tree with the 14
wind, and to return against the wind. When the wind continues to blow from the same quarter while the delinquent travels thirty, or six and thirty miles, if he be of a good constitution, he certainly survives. But what proves the most destructive is, that there is no dependence on the wind in that part of the world for any length of time.—There are no regular land winds; and the sea wind is not perceived there at all, the situation of the tree being at too great a distance, and surrounded by high mountains and uncultivated forests. Besides, the wind there never blows a fresh regular gale, but is commonly merely a current of light, soft breezes, which pass through the different openings of the adjoining mountains. It is also frequently difficult to determine from what part of the globe the wind really comes, as it is divided by various obstructions in its passage, which easily change the direction of the wind, and often totally destroy its effects.

I, therefore, impute the distant effects of the poison, in a great measure, to the 15
constant gentle winds in those parts, which have not power enough to disperse

the poisonous particles. If high winds were more frequent and durable there, they would certainly weaken very much, and even destroy the obnoxious effluvia of the poison; but without them, the air remains infected and pregnant with these poisonous vapours.

I am the more convinced of this, as the worthy ecclesiastick assured me that 16
a dead calm is always attended with the greatest danger, as there is a continual perspiration issuing from the tree, which is seen to rise and spread in the air, like the putrid steam of a marshy cavern.—[*London Mag.*]

Experiments made with the Gum *of the* Upas-Tree.

In the year 1776, in the month of February, I was present at the execution of 17
thirteen of the Emperor's concubines, at *Soura-Charte*, who were convicted of infidelity to the Emperor's bed. It was in the forenoon, about eleven o'clock, when the fair criminals were led into an open space within the walls of the Emperor's palace. There the judge passed sentence upon them, by which they are doomed to suffer death by a lancet poisoned with Upas. After this the Alcoran was presented to them, and they were, according to the law of their great prophet Mahomet, to acknowledge and to affirm by oath, that the charges brought against them, together with the sentence and their punishment, were fair and equitable. This they did, by laying their right hand upon the Alcoran, their left hands upon their breast, and their eyes lifted towards heaven; the judge then held the Alcoran to their lips, and they kissed it.

These ceremonies over, the executioner proceeded on his business in the fol- 18
lowing manner:—Thirteen posts, each about five feet high, had been previously erected. To these the delinquents were fastened, and their breasts stripped naked. In this situation they remained a short time in continual prayer, attended by several priests, until a signal was given by the judge to the executioner; on which the latter produced an instrument, much like the spring lancet used by Farriers for bleeding horses. With this instrument, it being poisoned with the gum of the Upas, the unhappy wretches were lanced in the middle of their breasts, and the operation was performed upon them all in less than two minutes.

My astonishment was raised to the highest degree, when I beheld the sudden 19
effects of that poison, for in about five minutes after they were lanced, they were taken with a *tremor*, attended with a *subsultus tendinum*, after which they died in the greatest agonies, crying out to God and Mahomet for mercy. In sixteen minutes by my watch, which I held in my hand, all the criminals were no more. Some hours after their death I observed their bodies full of livid spots, much like those of the *Petechiæ*, their faces swelled, their colour changed to a kind of blue, their eyes looked yellow, &c. &c.

About a fortnight after this, I had an opportunity of seeing such another execu- 20
tion at Samarang. Seven Malayans were executed there with the same instrument, and in the same manner, and I found the operation of the poison, and the spots in their bodies exactly the same.

These circumstances made me desirous to try an experiment with some ani- 21
mals, in order to be convinced of the real effects of this poison; and as I had then two young puppies, I thought them the fittest objects for my purpose. I accordingly procured with great difficulty some grains of Upas. I dissolved half a grain of that gum in a small quantity of arrack, and dipped a lancet into it. With this poisoned instrument I made an incision in the lower muscular part of the belly in one of the puppies. Three minutes after it received the wound the animal began to cry out most piteously, and ran as fast as possible from one corner of the room to the other. So it continued during six minutes, when all its strength being exhausted, it fell upon the ground, was taken with convulsions, and died in the eleventh minute. I repeated this experiment with two other puppies, with a cat, and a fowl, and found the operation of the poison in all of them the same, none of these animals survived above thirteen minutes.

I thought it necessary to try also the affect of the poison given inwardly, 22
which I did in the following manner. I dissolved a quarter of a grain of the gum in half an ounce of arrack, and made a dog of seven months old drink it. In seven minutes a retching ensued, and I observed, at the same time, that the animal was delirious, as it ran up and down the room, fell on the ground, and tumbled about; then it rose again, cryed out very loud, and in about half an hour after was seized with convulsions, and died. I opened the body, and found the stomach very much inflamed, as the intestines were in some parts, but not so much as the stomach. There was a small quantity of coagulated blood in the stomach, but I could discover no orifice from which it could have issued, and, therefore, supposed it to have been squeezed out of the lungs, by the animal's straining while it was vomiting.

From these experiments I have been convinced, that the gum of the Upas is 23
the most dangerous and most violent of all vegetable poisons; and I am apt to believe that it greatly contributes to the unhealthiness of that island. Nor is this the only evil attending it: hundreds of the natives of Java, as well as Europeans, are yearly destroyed and treacherously murdered by that poison, either internally or externally. Every man of quality or fashion has his dagger or other arms poisoned with it; and in times of war the Malayans poison the springs and other waters with it; by this treacherous practice the Dutch suffered greatly during the last war, as it occasioned the loss of half their army. For this reason, they have ever since kept fish in the springs of which they drink the water; and sentinels are placed near them, who inspect the waters every hour, to see whether the fish are alive. If they march with an army or body of troops into an enemy's country, they always carry live fish with them, which they throw into the water some hours before they venture to drink it, by which means they have been able to prevent their total destruction.

This account, I flatter myself, will satisfy the curiosity of my readers, and the 24
few facts which I have related will be considered as a certain proof of the existence of this pernicious tree, and its penetrating effects.

If it be asked why we have not yet any more satisfactory accounts of this 25
tree, I can only answer, that the object of most travellers to that part of the world consists more in commercial pursuits than in the study of Natural History and the advancement of Sciences. Besides, Java is so universally reputed an unhealthy island, that rich travellers seldom make any long stay in it, and others want money, and generally are too ignorant of the language to travel, in order to make enquiries. In future, those who visit this island will probably now be induced to make it an object of their researches, and will furnish us with a fuller description of this tree.

I will, therefore, only add, that there exists also a sort of Cajoe-Upas on the 26
coast of Macassar, the poison of which operates nearly in the same manner, but is not half so violent or malignant as that of Java, and of which I shall likewise give a more circumstantial account in a description of that island.—*London Magazine.*

SUPPLEMENT.

Addition to the note on Silene.

I saw a plant of the Dionæa Muscipula, Fly-trap of Venus, this day in the col- 1
lection of Mr. Boothby at Ashbourn-Hall, Derbyshire, Aug. 20th 1788. and on drawing a straw along the middle of the rib of the leaves as they lay upon the ground round the stem, each of them in about a second of time closed and doubled itself up, crossing the thorns over the opposite edge of the leaf, like the teeth of a spring rat-trap: of this plant I was favor'd with an elegant colour'd drawing, by Miss Maria Jacson of Tarporly, in Cheshire, a Lady who adds much botanical knowledge to many other elegant acquirements.

In the Apocynum Androsæmifolium, one kind of Dog's-bane, the anthers con- 2
verge over the nectaries, which consist of five glandular oval corpuscles surrounding the germ; and at the same time admit air to the nectaries at the interstice between each anther. But when a fly inserts its proboscis between these anthers to plunder the honey, they converge closer, and with such violence as to detain the fly, which thus generally perishes. This account was related to me by R. W. Darwin, Esq; of Elston, in Nottinghamshire, who show'd me the plant in flower, July 2d. 1788, with a fly thus held fast by the end of it's proboscis, and was well seen by a magnifying lens, and which in vain repeatedly struggled to disengage itself, till the converging anthers were separated by means of a pin: on some days he had observed that almost every flower of this elegant plant had a fly in it thus entangled; and a few weeks afterwards favor'd me with his further observations on this subject.

"My Apocynum is not yet out of flower. I have often visited it, and have fre- 3
quently found four or five flies, some alive and some dead, in it's flowers; they are generally caught by the trunk or proboscis, sometimes by the trunk and a leg; there is one at present only caught by a leg; I don't know that this plant sleeps, as the flowers remain open in the night, yet the flies frequently make their escape. In a plant of Mr. Ordino's, an ingenious gardener at Newark, who is possessed of a great collection of plants, I saw many flowers of an Apocynum with three dead flies in each; they are a thin-body'd fly, and rather less than the common

LP9

house-fly; but I have seen two or three other sorts of flies thus arrested by the plant. Aug. 12, 1788."

Addition to the note on Tropæolum.

In Sweden, a very curious phenomenon has been observed on certain flowers, by 4
M. Haggren, Lecturer in Natural History. One evening he perceived a faint flash of light repeatedly dart from a Marigold. Surprised at such an uncommon appearance, he resolved to examine it with attention; and to be assured that it was no deception of the eye, he placed a man near him, with orders to make a signal at the moment when he observed the light. They both saw it constantly at the same moment.

The light was most brilliant on Marigolds, of an orange or flame colour; but 5
scarcely visible on pale ones.

The flash was frequently seen on the same flower two or three times in quick 6
succession; but more commonly at intervals of several minutes; and when several flowers in the same place emitted their light together, it could be observed at a considerable distance.

This phænomenon was remarked in the months of July and August, at sun-set, 7
and for half an hour after, when the atmosphere was clear; but after a rainy day, or when the air was loaded with vapours, nothing of it was seen.

The following flowers emitted flashes, more or less vivid, in this order: 8

1 The Marigold, (*Calendula Officinalis*). 9

2 Garden Nasturtion, (*Tropæolum majus*). 10

3 Orange Lilly, (*Lilium bulbiferum*). 11

4 The Indian Pink, (*Tagetes patula et erecta*). 12

Sometimes it was also observed on the Sun-flowers, (*Helianthus annuus*). But 13
bright yellow, or flame colour, seemed in general necessary for the production of this light; for it was never seen on the flowers of any other colour.

To discover whether some little insects, or phosphoric worms, might not be the 14
cause of it, the flowers were carefully examined even with a microscope, without any such being found.

From the rapidity of the flash, and other circumstances, it may be conjectured, 15
that there is something of electricity in this phænomenon. It is well known, that when the *pistil* of a flower is impregnated, the *pollen* bursts away by its elasticity, with which electricity may be combined. But M. Haggren, after having observed the flash from the Orange-Lilly, the *antheræ* of which are a considerable space distant from the *petals*, found that the light proceeded from the *petals* only; whence he concludes, that this electric light is caused by the *pollen*, which in flying off, is scattered upon the *petals*.

[*Gen. Evening Post.*] 16

ADDITIONAL ILLUSTRATIONS

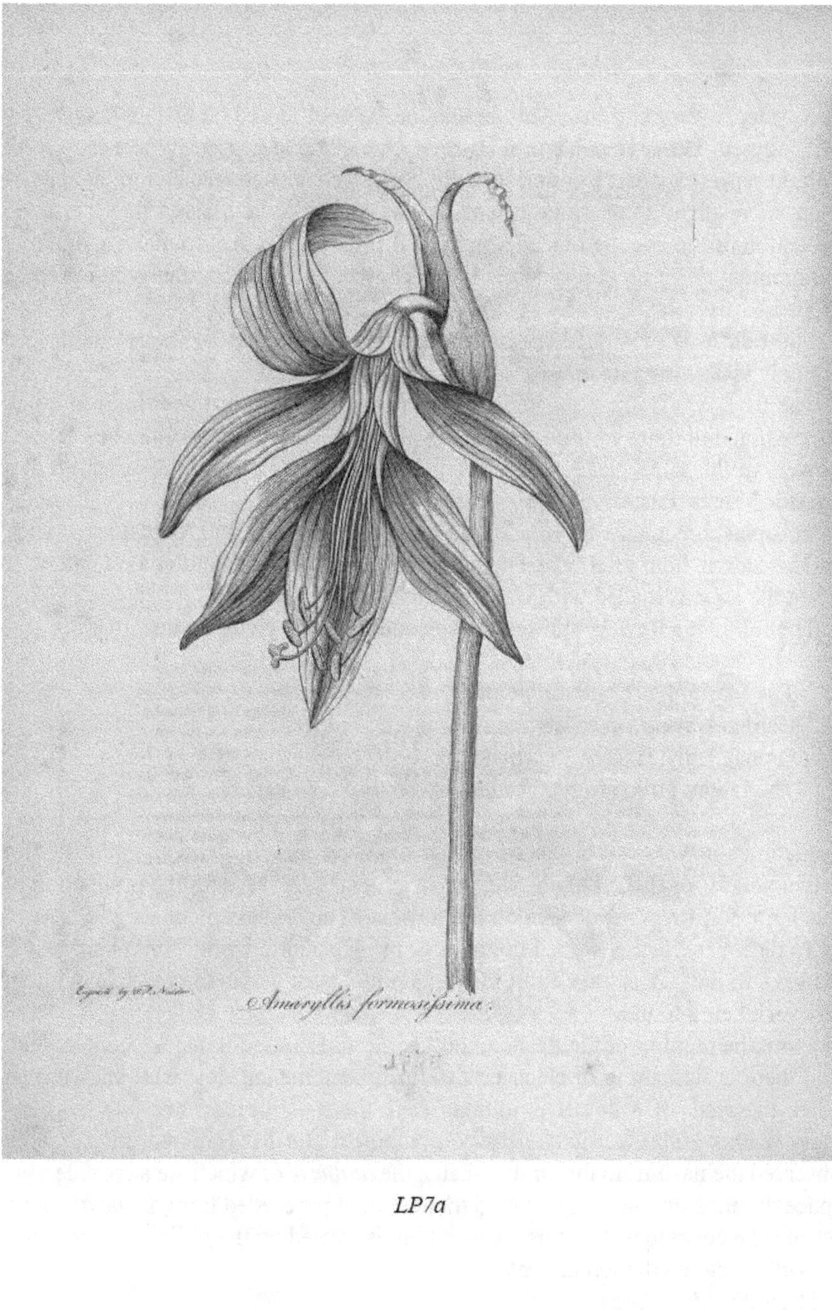

LP7a

The Tartarian Lamb.

Barometz.

Nightmare.

London, Published June 1 1791 by J. Johnson, St. Pauls Church Yard.

LP8a

LP8b

EXPLANATORY NOTES TO *THE LOVES OF THE PLANTS*

Title Page

Epigraph VIVUNT IN VENEREM ... ALNOQUE ASSIBILAT ALNUS Claudian (Claudius Claudianus) (c. 370–c. 404), *Epithalamium de nuptiis Honorii Augusti* 65–68, somewhat misquoted ("nemus omne per altum" for "omnisque vicissim", "faedera" for "foedera", "assibilat" for "adsibilat"): "The very leaves live for love and in his season every happy tree experiences love's power: palm bends down to mate with palm, poplar sighs its passion for poplar, plane whispers to plane, alder to alder" (trans. Maurice Platnauer).

Imprint *J. JACKSON* John Jackson (c. 1750–c. 1815), Lichfield printer and author.

Imprint Entered at Stationer's Hall The Stationers' Company was founded in 1403 to represent booksellers and related tradesmen; after being chartered by the crown in 1557, it gained an effective monopoly over the English book trade. Commercial printers were in most cases required to be members of this guild and to enter their publications in its register—both for a fee but in exchange for a perpetual copyright (later limited by the Copyright Act of 1709). In 1606, Stationer's Hall moved to its third and current location in Ave Maria Lane, near St. Paul's Cathedral.

Proem

2 CAMERA OBSCURA "Dark chamber"; originally, a room into which light would pass through a pinhole, projecting an inverted image of the exterior onto a whitened wall. Robert Hooke (English, 1635–1703) communicated "A Contrivance to Make the Picture of Any Thing Appear on a Wall, Cub-Board, or within a Picture-Frame, &c. in the Midst of a Light Room in the Day-Time", *Phil. Trans.* 3 (1668): 741–43; this used a much larger aperture, as well as "Reflecting and Refracting Glasses." Portable versions were introduced from the early eighteenth century. Figurative uses of the phrase date from mid-century.

3 P. Ovidius Naso ... Augustus Cæsar Publius Ovidius Naso or Ovid (43 BCE–17 CE) wrote the *Metamorphoses* around 8 CE. Its fifteen books conclude with the transformation of Julius Caesar, great-uncle and adoptive father of the emperor Augustus Caesar (63 BCE–14 CE), into an astral god.

Preface

3 the Dictionaire Botanique of M. Bulliard Pierre Bulliard (French, 1742–93), *Dictionnaire élémentaire de botanique* (Paris, 1783): pl. 2 ("Systeme sexuel de Linnaeus").

35 silicule Silicula, a short seed-pod.

35 *Shepherd's purse* *Thlaspi bursa-pastoris.*

35 silique Siliqua, a long seed-pod with a lengthwise partition.

35 *Wall-flower* *Cheiranthus.*

42 The Reader, who ... four Volumes Octavo See Preface 45–46.

43 thesis of Dr. Elmsgreen Johan Elmgren, "Termini Botanici", 22 June 1762, *Amœn. Acad.* 6.113 (1764): 217–46.

Canto I

1.1 Sylphs Here and in *EV*, Darwin draws on the Rosicrucian doctrine, a mystic order allegedly founded by Christian Rosenkreuz in the late fifteenth century (and treated facetiously in *The Rape of the Lock* [1712–14] by Alexander Pope [English, 1688–1744]). According to this system, a different kind of genius (local or governing spirit) inhabits each of the four elements: the Gnomes preside over earth; *Sylphs*, air; Nymphs, water; and Salamanders, fire.

1.3 fairy ... rings Fungal growth produces the circles on the ground known as "fairy rings", which contemporaries attributed also to lightning, insects or supernatural agency.

1.4 Gnomes See *LP* 1.1.

1.5 oaten reed Rustic pipe made from an oak stem, a pastoral convention.

1.10 vegetable Loves The Aristotelians used the term "vegetable soul" (*anima vegetabilis*) to refer to mere life without sentience. Until the discoveries in the late seventeenth century that Linnaeus would popularize in the eighteenth, "vegetable love" would also have suggested asexual reproduction, as some have argued of Andrew Marvell's "To His Coy Mistress." In *Zoonomia*, Darwin ventures farther beyond allegory, positing the difference between the vegetable and animal kingdoms as one of degree rather than of kind; plant reproduction is driven by the actual "sensation of love" rather than "mechanical attraction" (1.106).

1.11 Snow-drops *Galanthus*; see *LP* 4.103.

1.11 Harebels Also hare-bell, hairbell. Used by Darwin for *Hyacinthus* (see *LP* 1.177n), but by others for *Scillanutans* and, especially in Scotland, *Campanula.*

1.13 Violet *Viola*; see *LP* 1.213.

1.13 Primrose *Primula.*

1.15 Lily *Lilium* (see *LP* 4.103); compare *Gloriosa superba*, superb lily (*LP* 1.119n); *Nymphaea alba*, white water-lily (*LP* 2.163n); *Hemerocallis fulva*, day-lily (*LP* 2.163n); and *Convallaria*, lily of the valley (*LP* 4.103n).

1.16 Cowslips *Primula* (see *LP* 1.13); compare *Dodecatheon*, American cowslip (*LP* 1.61 and n).

1.17 Rose *Rosa*.

1.19 Woodbines *Lonicera*; see *LP* 1.211.

1.35 Graces In Greek mythology, the Graces (the sister goddesses Aglaia, Thalia and Euphrosyne) attended Aphrodite (Roman Venus); they were responsible for mental and physical endowments.

1.45n Ray John Ray (Wray) (English, 1627–1705), naturalist and theologian; author of the *Historia Plantarum* (3 vols., 1686–1704).

1.51 Collin Conventional name for a pastoral "swain", as in John Skelton's *Collyn Clout* (1522) and Edmund Spenser's *The Shepheardes Calender* (1579).

1.51n anther The part of the stamen (male floral organ) that contains pollen. The pistil (female floral organ) includes the style (filament) that terminates in the stigma.

1.51n Gloriosa, and Genista See *LP* 1.119n, 1.57n.

1.57 myrtle *Myrtus*; compare *Myrica cerifera*, candle-berry myrtle (*LP* 4.121).

1.57n stamens See *LP* 1.51n.

1.57n stigma ... pistil See *LP* 1.51n.

1.57n keel-leaf Later keel-petal, carina; one of the two lowermost petals, so called because they form the shape of a ship's prow.

1.57n Collinsonia, Gloriosa, Draba See *LP* 1.51n, 1.119n, 1.220n.

1.60n Marum ... Nepeta Darwin may have in mind *Teucrium marum*, cat-thyme; and *Nepeta cataria*, nep (whence "catnip") or cat-mint.

1.61n anthers See *LP* 1.51n.

1.61n Dr. Mead Richard Mead (English, 1673–1754), physician, collector and patron of the naturalist Mark Catesby, who named the genus *Meadia* (latter reclassified by Linnaeus) in his honor; see *The Natural History of Carolina, Florida and the Bahama Islands* (1729–47).

1.61n storgé Greek, parental love. Darwin will attempt to derive it in *The Temple of Nature* (2.92 and Additional Note 9) as an imitation of inanimate nature (e.g., a bird sits on its eggs after observing the sun warming them), or as a response to pleasurable sensations (i.e., incubating, seeing, touching and nursing one's young all stimulate the "glandular system ... into greater action") (40). Such affection is stronger in the mother, but "existed before animals were divided into sexes, and produced the beginning of sympathetic society" (40).

1.61n Ilex, and Gloriosa See *LP* 1.143n, 1.119n.

1.61n Cyclamen See *LP* 3.311.

1.61n style See *LP* 1.51n.

1.65n flax *Linum*; see *LP* 2.67.

1.65n sun-flower *Helianthus*; see *LP* 1.191.

1.65n Opulus *Viburnum opulus*; see *LP* 1.69n.

1.65n Perhaps all the products ... terraqueous globe See *AN* 24.

1.65n A.T. Bladh. Amaen. Acad. V. 7 Anders Johan Bladh (1748–1834), *Fundamenta entomologiae* (14 June 1767), *Amoen. acad.* 7.133 (1769): 129–59. The *Amoenitates academicae* was a collection of the 186 dissertations written under Linnaeus' tutelage.

1.68 Platonic love The *OED* dates this phrase, in the sense of nonsexual affection, to 1631.

1.69n flore pleno Double-flowered.

1.69n Stramoneum *Datura stramonium.*

1.69n Peonia *Paeonia.*

1.69n *Philos. Botan.* Linnaeus, *Philosophia Botanica* (Stockholm, 1751); rev. of *Fundamenta Botanica* (Amsterdam, 1736).

1.69n nectaries Sugar- or protein-secreting glands.

1.69n larkspur *Delphinium.*

1.69n columbine *Aquilegia.*

1.69n cyme A kind of inflorescence (i.e., an assemblage of flowers on a single plant) that is determinate (i.e., with its main axis terminating in a single flower that matures first, and with subsequent flowers maturing from the apex to the base).

1.69n gelder-rose *Viburnum.*

1.69n florets Small flowers making up an assemblage.

1.69n daisy *Bellis*; see *LP* 4.114.

1.69n feverfew *Matricaria.*

1.69n Plantago See *LP* 1.77.

1.69n calyx The "exterior Covering of a Flower"; that nearest the corol or "interior Covering of the Flower" is called the *perianth* (*SV* 6).

1.69n Hepatica *Anemone*; see *LP* 1.284.

1.69n poppies *Papaver*; see *LP* 2.268.

1.69n roses *Rosa*; see *LP* 1.17.

1.69n Nigella See *LP* 1.51n.

1.70 Eloisa Héloïse (c. 1098–1164), niece of Canon Fulbert of Notre-Dame de Paris and pupil of the theologian Pierre Abélard (1079–1142), with whom she fell in love. They married in secret following the birth of their son, Astrolabe. Abélard was castrated at Fulbert's urging and became a monk of Saint-Denis, while Héloïse became a nun of Argenteuil. Later, Abélard founded and made Héloïse abbess of the monastery Le Paraclet, where they were both interred. Their love-letters, only seven of which survive, have long been celebrated. These were available in various forms during the eighteenth century, including John Hughes' English translation of a French compilation (London, 1713), Richard Rawlinson's edition of the original Latin (London, 1718) and Joseph Berington's *History of the Lives of Abeillard and Heloisa* (Birmingham, 1787). Among those works inspired by the story (and themselves inspiring many successors) were the verse-epistle *Eloisa to Abelard* (1717) by Alexander Pope (see *LP* 1.1), who consulted Hughes; and Jean-Jacques Rousseau's novel *Julie, ou nouvelle Héloïse* (1761).

1.73n nettle *Urtica*; see *LP* 3.191.

1.73n resins ... Venice turpentine The viscous exudate or resin of coniferous trees was known simply as turpentine, with the qualifier indicating the kind of tree. Its volatile portion, called oil or spirit of turpentine, can be separated by distillation. See also *LP* 3.184n.

1.73n larch *Pinus larix*.

1.73n Sandarach Also sandarac, sandarack; its resinous form is used in varnish, its pulverized form in a blotting-paper treatment called pounce.

1.73n common juniper *Juniperus communis*.

1.73n incense The precise source of frankincense or olibanum, an aromatic sap with ancient ritual and medicinal uses, was disputed in eighteenth-century Britain; Darwin accepts the prevailing identification as what Linnaeus called *Juniperus lycia*. It is now identified as the genus *Boswellia* (family Burseraceae, order Sapindales), which was first described and named by William Roxburgh and reported by Henry Thomas Colebrooke in *Asiatick Researches* 9 (1809): 377–82.

1.73n Egyptian mummies The scholarly and popular fascination with Egypt was millennia old. Eighteenth-century discourses on mummies included those by Guillame-François Rouelle (*Hist. de l'Acad. roy.*, 1750) and Anne Claude comte de Caylus (*Recueil d'antiquités*, 1752–67?). Notable specimens included those donated by George Townshend to Cambridge University in 1743, and by William Lethieullier to the British Museum from 1755.

1.73n Cedar Also Bermuda cedar; *Juniperus bermudiana*.

1.73 black-lead Graphite, a form of carbon.

1.73n Miln's Bot. Dict. Colin Milne (Scottish, c. 1743–1815), *A Botanical Dictionary* (London, 1770).

1.75n cochineal insect Cactus-feeding insect found primarily in Mexico; the dried bodies of the female are used to make the red dye called cochineal and pigment called carmine.

1.77n bractes Small, modified leaves growing directly beneath the calyx (see *LP* 1.69n) of a flower. Collectively, *chaff*; in grasses, *glume*.

1.77n spike A kind of inflorescence that is indeterminate (i.e., with flowers maturing from the base to the apex; compare note to *cyme*, *LP* 1.69n) and that consists of sessile (i.e., stalkless) flowers alternating on a single axis.

1.77n Dianthus See *LP* 4.207.

1.77n ament Amentum or catkin; a unisexual spike that falls off as a unit after flowering.

1.77n salix rosea Not in Linnaeus; what was commonly called the *rose-willow* was classed as *Salix helix* (now *Salix purpurea*).

1.77n Alcea See *LP* 1.69.

1.81–84 So hapless Desdemona ... his sooty breast *Othello* 1.3.127–68.

1.86 ambrosial "Divinely fragrant" (*OED*), as of ambrosia, the legendary food of the Greek gods.

1.86n birch-trees *Betula*.

1.86n A. T. Bladh. Amæn. Acad. V. 7 See *LP* 1.65n.

1.87 purple An ancient imperial color.

1.88 furze *Ulex*.

1.97n artichokes *Cynara*.

1.97n globe-thistles *Echinops*.

1.97n Discourse on irritability of plants. Dodsley Giovambatista dal Covolo (Italian, 1739–68), *A Discourse Concerning the Irritability of Some Flowers*, trans. Benjamin Stillingfleet (London: J[ames] Dodsley, 1767).

1.101 Eolian Lyre Also Aeolian lyre, Aeolian harp; from Aeolus, the Greek god of the winds. A zither played by the wind, usually when placed in a window. This ancient instrument was reinvented by the German Jesuit priest Athanasius Kircher, who described it in his *Musurgia universalis* (1650). It was often discussed for its relation to the human mind, both literal (James Thomson, *The Castle of Indolence*, 1748; James Beattie, *Essay on Poetry and Music*, 1762) and metaphorical (Samuel Taylor Coleridge, "Effusion XXXV", in *Poems on Various Subjects*, 1796, rev. as "The Eolian Harp" for *Sybilline Leaves*, 1817; Coleridge, "Dejection: An Ode", *Morning Post* 4 October 1802, rev. for *Sybilline Leaves*, 1817; Percy Bysshe Shelley, *Ode to the West Wind*, 1820).

1.103 Zephyr The west wind.

1.106 Diana The (Roman) goddess of the moon, chastity and the hunt; Greek Artemis.

1.108n corol See *LP* 1.69n.

1.117 Hours Horae; mythic female beings who preside over the changing of the seasons.

1.119n Lithrum *Lythrum*.

1.119n Lychnis See *LP* 1.108.

1.119n Genista See *LP* 1.57.

1.121 bark Crust; Darwin alludes to Pope's *Odyssey*, "O'er thy smooth skin a bark of wrinkles spread" (13.457).

1.125 Ninon Ninon (Anne) de Lenclos (Lanclos) (French, 1620–1705), libertine author of *La Coquette vengée* (1659). She entertained some of the most eminent men of France as lovers and as guests of her esteemed Parisian salon. Late in life, she was reunited with Charles de Villiers, who (unbeknownst to him) was her illegitimate son by the Marquis of Jersey. He fell in love with her and, when she revealed her identity, stabbed himself to death.

1.131 fell Fierce.

1.131n Cucubulus *Cucubalus*.

1.131n The last professor Linneus ... supplementum plantarum Carl von Linné the Younger (Swedish, 1741–81), *Supplementum plantarum* (Brunswick, 1781). It incorporated the thirteenth edition of his father's *Systema vegetabilium* (Göttingen and Gotha: Johan Anders Murray, 1774), the sixth of the *Genera plantarum* (Stockholm, 1764; 1st ed., New York, 1737) and the second of the *Species plantarum* (Stockholm, 1762; 1st ed., Stockholm, 1753). *A System of Vegetables* (*SV*) (Lichfield, 1783[–85]) drew on the thirteenth *Systema* and other parts of the *Supplementum*. Darwin was sole translator for *SV*, though he consulted the two other members of the Lichfield Botanical Society, Brooke Boothby (English,

1744–1824) and William Jackson (English, 1734/35–98); as well as some forty eminent botanists, including the younger Linnaeus himself. Darwin's subsequent translation *Families of Plants* (Lichfield, 1787) also drew on the *Supplementum*, but using the seventh *Genera* (Frankfurt, 1778).

1.131n Arum Muscivorum See *LP* 4.189. Now classed as *Helicodiceros muscivorus*, dead horse arum lily.

1.131n Dypsacum *Dipsacus*. See *LP* 1.333.

1.131n Drosera See *LP* 1.199.

1.131n Cerea See *LP* 4.15.

1.133 bravoes "Daring villain[s] ... hired soldier[s] or assassin[s] ... reckless desperado[es]" (*OED*).

1.138 Sirens In the *Odyssey* (12.39 ff.), the sirens were part-woman, part-bird creatures whose singing lured sailors toward the rocks.

1.143n barberry *Berberis*.

1.143n hawthorn *Crataegus*.

1.143n gooseberry *Ribes*.

1.143n gorse *Ulex*; see *LP* 1.88n.

1.143n Needwood-forest This Staffordshire forest covered nearly 10,000 acres; plans to partition and cultivate it were set in motion in 1656, quashed by Charles II and revived in the late eighteenth century. Francis Noel Clarke Mundy (English, 1738/39–1815) published a five-part poem, *Needwood Forest* (Lichfield, 1776), opposing deforestation; an excerpt was appended to *3LP1791+* (see textual notes). The Mundy volume also included Darwin's "Address to Swilcar Oak", Anna Seward's "A Rural Coronation", a "Sonnet" by Brooke Boothby, and Erasmus Darwin, Jr.'s "On Mr. Mundy's *Needwood Forest.*" Mundy's "Address to the River Derwent" was prepended to *3EV1795+* (see textual notes).

1.143n Meadia See *LP* 1.61.

1.143n Mancinella See *LP* 3.188.

1.143n bird-lime "A glutinous substance spread upon twigs, by which birds may be caught and held fast" (*OED*).

1.147 Moore Hero of the bawdy broadside *A True Relation of the Dreadful Combate between More of More-Hall and the Dragon of Wantley* (1685); clad in a spiked suit of armor, he braves a dragon that has been terrorizing a town. He first conceals himself in a well and hits the dragon in the mouth as it prepares to drink; later he kills it with a "kick on the Arse." The ballad was reprinted as the lead entry in the 1699 edition of Henry Playford's *Wit and Mirth; or, Pills to Purge Melancholy*; in expurgated form, with commentary, in Thomas Percy's *Reliques of Ancient Poetry* (1765); and throughout the eighteenth century. Henry Carey adapted it as *The Dragon of Wantley, A Burlesque Opera*, performed at the Theatre Royal in Covent Garden with music by John Fredericke Lampe in 1737. The legend may have originated with a sixteenth-century lawsuit conducted by the real More of Wantley (Warncliffe), South Yorkshire.

1.153n Needwood's wide domains See *LP* 1.143n.

1.157n Adansonia ... Adanson ... Voyage to Senegal Linnaeus does not place *Adansonia* and *Kleinhovia* in the same order, though modern taxonomists do. Michel Adanson (French, 1727–1806), *Histoire naturelle du Sénégal...pendant les années 1749, 50, 51, 52 & 53* (Paris, 1757); trans. *A Voyage to Senegal, the Isle of Goree, and the River Gambia* (London, 1759).

1.165 Thalestris Amazonian queen; attended by three hundred warrior women in martial habit, she visited Alexander the Great at Hyrcania to conceive a valiant child. She found Alexander less than impressive at first sight but spent thirteen days with him and achieved her aim; after her departure, he fell into dissipation and effeminacy among the Persians. The story appears in Diodorus Siculus' *Historical Library* and in Quintus Curtius' *History of Alexander the Great*; John Weston retells it in his stage-farce *The Amazon Queen* (1667); and Pope references it in *The Rape of the Lock* (1714), where the heroine Belinda's irascible handmaid is named Thalestris.

1.168 Bellona Roman war-goddess (Greek Duellona, Enyo), sometimes identified as the sister or wife of Mars.

1.177 Hyacinth *Hyacinthus*; see *LP* 1.11.

1.177n Nymphaea See *LP* 2.163; *Nymphaea nelumbo* is now classed as *Nelumbo nucifera*.

1.177n Mr. Ferber Johann Jakob Ferber (Swedish, 1743–90), *Prolepsis plantarum* (12 June 1763), *Amoen. acad.* 6.120 (1763): 365–83.

1.177n Mariotte Edmé Mariotte (French, c. 1620–84), physicist; he made his presentation to the Académie royale des sciences concerning the tulip on 17 November 1677.

1.177n Hepatica *Anemone hepatica*; see *LP* 1.284.

1.177n Osmunda See *LP* 1.93; *Osmunda lunaria* is now classed as *Botrychium lunaria*.

1.181 Dormouse Long-hibernating rodent; from Latin *dormire*, to sleep.

1.184n Miller's Dict. Philip Miller (English? 1691–1771), *The Gardeners Dictionary* (London, 1732).

1.184n deciduous Shedding their leaves or other parts at a certain season or developmental stage.

1.184n ash *Fraxinus*.

1.184n Phil. Bot. See *LP* 1.69n.

1.184n Lin. Spec. Plant. See *LP* 1.131n.

1.189 Georgian star Or Georgian planet, *Georgium Sidus*; controversially named by Sir Frederick William Herschel (German, 1738–1822), who discovered it in 1781, after his patron George III. His fellow astronomer Johann Elert Bode proposed the name Uranus instead, which did not gain general acceptance until 1850.

1.191 Helianthus See *LP* 1.65n.

1.191n Chondrilla See *LP* 1.97.

1.191n nutation ... Hales veg. stat. Bending; in the botanical sense, Stephen Hales' (English, 1677–1761) usage in *Vegetable Staticks* (London, 1727) predates the first citation, from Darwin, in the *OED*. Hales explains, "the side of the stem next the Sun perspiring most, it shrinks" (39).

1.191n Mimosa See *LP* 1.267.

1.192 Dervice Or dervish; a "Muslim friar, who has taken vows of poverty and an austere life"; known for "fantastic practices as dancing or whirling" (*OED*).

1.194n mistleto *Viscum*; see *LP* 1.226.

1.194n clevers Also cleavers, clivers, goose-grass, goose-grease; *Galium aparine*.

1.194n burdock *Arctium*.

1.194n houndstongue *Cynoglossum*.

1.194n hawthorn See *LP* 1.143n.

1.194n juniper See *LP* 1.73n.

1.194n Oats *Avena*; see *LP* 4.73.

1.194n Geranium See *LP* 1.65n.

1.194n Impatiens See *LP* 3.131.

1.194n Zostera See *LP* 1.232.

1.194n Cassia See *LP* 3.345.

1.194n Carlina See *LP* 2.7.

1.199 Drosera See *LP* 1.131n.

1.199n Silene See *LP* 1.131.

1.199n Mr. Wheatly ... Cateaton-street Not identified, but this is presumably the Wheatly to whom Darwin refers in *Letters* 79–11 and 88–16. Cateaton-Street, Cheapside, is now Gresham Street.

1.199n muscipula veneris *Dionaea muscipula*, Venus fly-trap; see *LP* 1.131n.

1.200 rushy Rushes are marsh plants with hollow or pith-filled stems, especially *Juncus*.

1.208 zone Girdle, belt or band; often signifies the power conferred by righteousness or by an enticing sexual restraint, as Aphrodite's "golden zone" that Hera's (Roman Juno's) "swelling bosom bound" (Pope's *Iliad* 14.210).

1.211 Lonicera See *LP* 1.19.

1.211n cornucopiae, or horn of plenty In Greek mythology, while the infant Zeus (Roman Jupiter, Jove) lay hidden on Mount Ida from his father Cronos (Roman Saturn), the daughters of the Cretan king Melisseus fed him with the milk of the she-goat Amalthea; after she died, Zeus honored her by breaking off one of her horns and bestowing on it the power of becoming filled with whatever food or drink its possessor wished. Another version of the legend holds that Hercules strove with the river-god Achelous, who assumed the form of a bull, to win the favor of the maiden Dejanira; Hercules threw Achelous and tore off one of his horns, which the Naiades filled with flowers.

1.211n Helleborus See *LP* 2.199.

1.211n Aquilegia See *LP* 1.69n.

1.211n Sphinx Genus first identified by Linnaeus in *Systema Naturae*, 10th ed. (Stockholm, 1758; 1st ed., Leiden, 1735).

1.211n Delphinium See *LP* 1.69n.

1.211n Rubia See *LP* 1.341.

1.211n Conferva polymorpha See *LP* 4.366.

1.212 vermil Also vermeil, vermillion; bright scarlet.

1.213 pansied First *OED* citation; pansies are *Viola* (see *LP* 1.13).

1.216 Horn of Plenty See *LP* 1.211n.

1.219 Tenerif Also Tenerife, Teneriffe; a complex of stratovolcanoes that constitutes the largest of the Canary Islands, Spain, off the northwestern coast of Africa. It had last erupted in 1704–06.

1.220n Gloriosa See *LP* 1.119.

1.220n antiscorbutic Acting against scurvy.

1.220n mustard *Sinapis*.

1.220n watercress *Sisymbrium nasturtium*.

1.220n cabbage *Brassica*.

1.220n turnep *Brassica rapa*.

1.226 Visca See *LP* 1.194n.

1.226n (Phil. transact.) George [Mackenzie, 1st] Earl of Cromertie [Cromarty] (Scottish, 1630–1714), "An Account of the Mosses in Scotland", *Phil. Trans.* 27 (1710–12): 296–301.

1.229 ether Poetically, "the upper regions of space beyond the clouds" or "the medium supposed to fill" them, "as the air fills the lower regions" (*OED*). For a more technical consideration, see *EV* 1.105 ff.

1.232n leagues A league is a highly irregular measure of distance; in England it signifies approximately 3 statute miles.

1.232n Scyllœa ... Barbut's genera vermia James Barbut (1776?–91 or 99), *The Genera Vermium Exemplified by Various Specimens of the Animals Contained in the Orders of the Intestina et Mollusca Linnaei* (London, 1783), cites the thirteenth edition of Linnaeus' *Systema Naturae* on the mollusk Scyllæa. In giving "sea onion" as its common name, however, Barbut seems to confuse it with the plant *Scilla*.

1.232n Needham John Turberville Needham (1713–81), the first English-born Roman Catholic priest to join the Royal Society. He recounts the experiments Darwin describes in his "Letter ... Concerning Certain Chalky Tubulous Concretions, Called Malm", *Phil. Trans.* 42 (1742–43): 634–41.

1.232n eolopile Or aeolopile; "an experimental or educational device comprising a container which can be filled with steam or other heated vapour which escapes through one or more narrow apertures with sufficient force to cause the container to rotate. Often considered the first steam engine" (*OED*).

1.232n Spallanzani ... dissertations Lazzaro Spallanzani (Italian, 1729–99), *Dissertazioni di fisica animale e vegetabile* (Modena, 1780), trans. Thomas Beddoes as *Dissertations Relative to the Natural History of Animals and Vegetables* (London, 1784).

1.232n Vallisneria See *LP* 1.361.

1.233 sea-weed *Fucus*; see *LP* 4.161.

1.245 conch Mollusk shell associated in classical mythology with the god Triton, who blew into it to control the sea.

1.250n decumbent "Lying or trailing upon the ground, but with the extremity ascending: applied to stems, branches, etc." (*OED*); first entry for this sense.

1.250n Sir Hans Sloane Irish, 1660–1753; *Phil. Trans.... . Abridg'd* 2 (London, 1705), no. 26 and fig. 160, p. 646.

1.250n Dr. Hunter ... Terra of Evelyn John Evelyn (English, 1620–1706), *Silva ... to Which Is Added the Terra*, ed. Alexander Hunter (Scottish, c. 1729–1809), 2 vols. (York, 1786), opp. 2.48. See Illustration LP7b.

1.250n blow-pipe "A tube through which a current of air (or other gas) is blown into a flame to increase the heat, for the purpose of fusing metals, etc.; especially employed in chemical experiments, analysis, etc." (*OED*).

1.250n camphor Volatile oil, $C_{10}H_{16}O$, prepared from the camphor or camphire tree (*Laurus camphora*; see *LP* 3.184n).

1.250n capillary attraction A "capillary tube" was one having the thinness of a hair. It was understood that the particles of a liquid in such a tube would be attracted to the walls of the tube and to each other (the "attraction of cohesion"), giving the liquid the ability (observed by da Vinci, Newton and others) to rise of its own accord. This phenomenon of "capillary attraction" (later "capillary action") was quantified by Thomas Young and by Pierre-Simon, marquis de Laplace in 1805.

1.253 grey coral moss Perhaps *Lichen* (see *LP* 1.315).

1.253 thyme *Thymus*.

1.254 rime Hoar-frost, frozen dew.

1.267 Mimosa The "sensitive plant" stood at the center of long-running debates as to whether the difference between vegetable and animal life was one of kind or of degree. Unlike Darwin, John Locke had inclined to the former, attributing the motions of the mimosa to "bare mechanism" rather than sentience (*Essay Concerning Human Understanding*, 1689). Even for some of Darwin's contemporaries, the mimosa had a merely analogical relationship to human beings; e.g., as a metaphor for the sexual organs of male aristocrats in James Perry's bawdy satire *Mimosa, or The Sensitive Plant* (London, 1779); or as a simile for feminine "delicacy" in Rev. John Bennett's *Letters to a Young Lady* (London, 1789). Darwin's account influenced Percy Shelley's "The Sensitive Plant" (1820).

1.267n note on vegetable respiration See *AN* 37.

1.275–78 Veil'd, with gay ... of her Lord Conventional Orientalism; the mimosa had been found principally in the Americas, the West Indies and parts of Africa, but not the Levant.

1.280 liquid silver ... glassy tower The mercury (quicksilver) barometer was invented by Evangelista Torricelli (Italian, 1608–47) in 1643; the mercury thermometer, by Daniel Gabriel Fahrenheit (Polish/Dutch, 1686–1736) in 1714.

1.282 librations "Motion like that of the beam of a balance oscillating upon its pivot" (*OED*).

1.284 Anemone See *LP* 1.69n.

1.284n Pliny says ... whence its name Pliny the Elder (23–79 CE), *Natural History* (77 CE), book 21, ch. 94. The Greek *anemone* translates as "daughter of the wind."

1.284n calix Or calyx; see *LP* 1.69n.

1.284n ranunculus See *LP* 1.69n.

1.284n melliferous Honey-yielding.

1.288n There is a wonderful ... Stillingfleet ... appearance in Greece Benjamin Stillingfleet's (English, 1702–71) *Miscellaneous Tracts Relating to Natural History, Husbandry and Physick* (London, 1759) contains his annotated translations of selected papers from the *Amoen. acad.* Darwin's remark, including the reference to Linnaeus, closely follows one of Stillingfleet's notes to "On the Foliation of Trees", *Miscellaneous Tracts*, 2nd ed. (London, 1762), trans. of *Vernatio arborum* by Haral Barck, *Amoen. acad.* 46 (5 May 1753): 363–76.

1.288n birds of passage Migratory birds.

1.288n wood anemone *Anemone sylvestris.*

1.288n marsh mary-gold, Caltha See *LP* 1.69n.

1.288n rose and the nightingale For its knowledge of the Arabic, Persian and Sanskrit languages and literature, eighteenth-century England was heavily indebted to the work of Sir William Jones (English, 1746–94), orientalist and judge. His *Grammar of the Persian Language* (London, 1771) includes a translation of the fable "The Gardener and the Nightingale", which elaborates the theme Darwin identifies. Sir William Ouseley's later *Persian Miscellanies* (London, 1795) gives additional examples from Sanā'ī (d. 1131?), Sa'dī (c. 1213–91) and Jāmī (1414–92).

1.288n potatoes, onions *Solanum* (see *LP* 1.61n), *Allium* (see 1.86n).

1.288n malt "Barley or other grain prepared for brewing, distilling or vinegar-making, esp. by steeping, germinating and kiln-drying" (*OED*).

1.288n Fordyce on agriculture George Fordyce (Scottish, 1736–1802), physician; *Elements of Agriculture and Vegetation* (Edinburgh, 1765; London, 1771).

1.288n Kalm's travels Pehr Kalm (Swedish, 1716–79), botanist; *Travels into North America*, trans. John Reinhold Foster, 3 vols. (London, 1770–71).

1.288n Kenedy on Gardening John Kennedy, agriculturalist and gardener to Sir Thomas Gascoigne; *A Treatise upon Planting, Gardening and the Management of the Hot-House* (York, 1776; 2 vols., London, 1777).

1.288n Dr. Walker's John Walker (Scottish, 1731–1803), minister and natural historian; "Experiments on the Motion of Sap in Trees", *Edinb. Trans.* 1 (1788): 3–40. Walker concludes that "the more a tree is replenished with sap, the less degree of heat is necessary to make it bleed" (19).

1.292 pencil Paintbrush.

1.300 sliding shell Unclear; perhaps the scallop-shell on which Zephyrus (see *LP* 1.103) wafted Venus ashore at Cyprus (as in Sandro Botticelli's famous painting *The Birth of Venus*, c. 1485), a conch-shell that sounds when blown, or a shell-shaped vessel generally.

1.306 calash "A woman's hood made of silk, supported with whalebone or cane hoops, and projecting beyond the face" (*OED*).

1.310 Landau Four-wheeled carriage with a retractable rear and removable front hood, named after the town in Germany where it was invented.

1.313 Snowden Or Snowdon, mountain in northern Wales; the highest point in Wales and England.

1.314 Conway Or Conwy; a county borough, a town and, here, a river in north-western Wales.

1.315n In this manner ... subterraneous fires See, principally, the "Geological Recapitulation" to *AN* 24.

1.318 flinty See *EV* 1.229n, *AN* 19.

1.323 Hymen Greek god of marriage, depicted as a male youth carrying a torch and bridal veil.

1.327 front Forehead.

1.327 Sirius The dog-star, chief in the constellation Canis Major and the brightest in the night sky; one origin of the sixteenth-century expression "dog days" in that, during the summer in the northern hemisphere, it rises with the sun and was believed by the ancients to compound its effects.

1.333 Dypsaca See *LP* 1.131n.

1.333n Tillandsia See *LP* 1.226n.

1.333n Dampier's voyage to Campeachy William Dampier (English, 1651–1715), explorer and pirate; *Voyages and Descriptions*, vol. 2 (London, 1699; vol. 1 apparently *A New Voyage Round the World*, London, 1697), Part 2 ("Two Voyages to Campeachy"), ch. 2. Dampier visited the Bay of Campeche, southeastern Mexico, in 1675–76.

1.333n Dr. Sloane See *LP* 1.250n; *A Voyage to the Islands Madera, Barbados, Nieves, S. Christophers and Jamaica, with the Natural History...of the Last...* (London, 1707), vol. 1, book 1, ch. 16, sect. 7. Sloane traveled to Jamaica as personal physician to its governor, the Duke of Albermale, in 1687; the specimens he brought back with him fifteen months later would form one of the three core collections of the British Museum, which opened in 1759.

1.333n aloe Sloane may describe what Linnaeus would classify as *Agave americana*, known popularly but misleadingly as the great or American aloe.

1.333n Banana *Musa*.

1.333n Burm. Zeyl. 42. 17 Johannes Burmann (Dutch 1707–80), botanist and physician; *Thesaurus zeylanicus* (Amsterdam, 1737); p. 42, pl. 17 depicts the *Bandura zeylanica*, which Linnaeus would rename *Nepenthes distillatoria*.

1.336 Naiads In Greek mythology, freshwater spirits.

1.341n Belchier John Belchier (English, c. 1706–85), F.R.S., surgeon of Guy's Hospital, London; "An Account of the Bones of Animals Being Changed to a Red Color by Aliment Only", *Phil. Trans.* 39 (1736): 287–88.

1.341n Comment. de rebus. Lipsiae, Vol... . *Commentarii de rebus in scientia naturali et medicina gestis* (Lipsiae [Leipzig], vol. 1 (1752): 519–21);

on Johann Benjamin Boehmer, *Radicis rubiae tinctorum effectus in corpore animali* (Leipzig, 1751).

1.342 vermil See *LP* 1.212.

1.346 woof Or weft, a "woven fabric"; more technically, the "threads that cross from side to side of a web, at right angles to the warp" (*OED*).

1.349–58 So when MEDEA ... round his brow *Aeson*, father of Jason, was king of Thessaly (eastern Greece); he abdicated to his brother, Pelias. When Jason came of age, rather than allow him to claim the crown, Pelias encouraged him to retrieve the golden fleece that resided at *Colchis* (modern-day Georgia). Medea, daughter of the Colchian king Aetes, used her sorcery to help Jason accomplish his quest and, later, to restore the infirm Aeson to the vigor of youth. For the grim epilogue, see *LP* 3.135–78.

1.361n Rhone River running from the Swiss Alps, through Lake Geneva and southeastern France, to the Mediterranean Sea.

1.361n Families of Plants See *LP* 1.131n.

1.369 shelving Gradually sloping.

1.373n duplicatures Folds.

1.373n Patavia Padua, city in northeastern Italy, known for its mineral and mud baths.

1.373n ulva labyrinthi formis Linnaeus' *Ulva labyrinthiformis* is now classed as *Spirulina labyrinthiformis*.

1.373n The air contained ... sometimes less pure Joseph Priestley (English, 1733–1804), radical theologian, political theorist and scientist; "Of the State of Air Confined in the Bladders of Sea Weed", *Experiments and Observations Relating to Various Branches of Natural Philosophy* (London, 1779), sect. 29; building on his *Experiments and Observations on Different Kinds of Air* (London, 1774). Of his many achievements, Priestley's contributions to the field that came to be called pneumatic chemistry—air-science—are among the most significant. The ancients had posited "air" as a single element; modern scientists discovered multiple "airs" with distinct properties. These airs (which came to be called by the names of gases, and were reconceptualized in the process) were in addition to the breathable atmospheric mixture known as "*common air*" or simply "air." They included, e.g., "inflammable air" (later hydrogen); "*dephlogisticated*", "pure" or "vital air" (later oxygen); and "nitrous air" (later nitric oxide, NO). Priestley measures the "*purity*" or "goodness" (i.e., breathability) of the air contained inside certain aquatic plants and compares it to that of common air. His method, which he employed throughout his career as preferable to testing on flames or on live animals, was to combine equal parts of the air in question and nitrous air trapped over water. The smaller the volume of the combination, the better the air in question (i.e., the less rich in an impurity called phlogiston). For example, air extracted from seaweed was judged better than common air, that from hemlock worse.

1.373n Monro. Physiol. of Fish Alexander Monro secundus (Scottish, 1733–1817), *The Structure and Physiology of Fishes Explained* (Edinburgh, 1785).

1.373n Ketmia *Hibiscus*.

1.373n cotyledon One of the "*Glandules* dispersed up and down the uttermost *Membrane* of the *Fœtus*, called *Chorion*; which separate *Nutritious* juice from the Womb to nourish the *Fetus*; but this is found onely in some Animals; the *Placenta* in the Womb supplies their Place in Women" (Steven Blankaart, *The Physical Dictionary* (London, 1648)).

1.373n umbelliferous Bearing umbels, inflorescences whose pedicels (flower stalks) are all attached at one point.

1.373n pith "The soft internal tissue of a plant part ... or animal structure; *esp.* the spongy core of a feather-shaft" (*OED*).

1.373n Some of these ... lungs in birds John Hunter (Scottish, 1728–93), "An Account of Certain Receptacles of Air, in Birds, Which Communicate with the Lungs, and Are Lodged both among the Fleshy Parts and in the Hollow Bones of Those Animals", *Phil. Trans.* 64 (1774): 205–13.

1.373n Mr. Day John Day, a Yarmouth carpenter, designed an air-chamber that would equip an ordinary vessel for diving. The fifty-ton sloop *Maria* was purchased on his behalf by one Charles Blake and fitted out by a Plymouth shipwright named Hunn. Day launched his experiment on 20 June 1774 and drowned when the *Maria* promptly sank to the bottom of Plymouth Sound.

1.375 Cyprian Queen Venus; see *LP* 1.300.

1.376 halcyons Kingfishers. According to legend, they breed during the winter solstice in a nest afloat the sea, which they charm to stillness; hence the expression *halcyon days*.

1.380 Galatea Sea-nymph, lover of the naiad (see *LP* 1.336) Acis. Killed by the jealous Cyclops Polyphemus, Acis was turned into a river by Galatea, who then joined her sister-nymphs under the sea.

1.388 Tritons Sea-deities or mermen.

1.393 Dove River in the Peak District of central England, between Staffordshire and Derbyshire.

1.393n wheat *Triticum*.

1.393n Ostiahs Ostyaks, now called Khanty; indigenous people principally of the Ob River basin, central Russia.

1.393n History of Russia ... 1780 Johann Gottlieb Georgi (German, 1738–1802), *Russia: or, a Compleat Historical Account of All the Nations Which Compose That Empire*, 4 vols. (London: J. Nichols et al., 1780–83): 1.173–201 ("The Ostiaks").

1.393n Agaricus officin. *Agaricus officinalis* is neither in Linnaeus nor in Georgi.

1.393n fungus vinosus So named by the Bristol surgeon James Ford, "An Account of the Success of Agaric, and the *Fungus vinosus*, in Amputations", *Phil. Trans.* 49 (1755–56): 93–95, in that it "grows in great abundance on the casks, walls, &c. in wine-vaults, and ... the coopers here apply [it] to their wounds, whenever they cut themselves."

1.393n esculent Edible.

1.400 Eurus The east wind.

1.402 Hesper The evening star.

1.403 northern dawn Literal translation of *aurora borealis*, coined by the French astronomer Pierre Gassendi in 1621 (slightly altering Galileo's *boreale aurora*). A major display was observed throughout Europe on the evening of 6 March 1716. The astronomer Edmond Halley (English, 1656–1742) attributed such phenomena to a magnetic "Effluvia" (naturally luminous or made so by friction with surrounding matter) that surrounded the earth and flowed toward its poles; *Phil. Trans.* 29 (1714–16): 406–28. The French physician Jean-Jacques Dortous de Mairan, who cataloged aurorae dating back to the sixth century in his *Traité physique et historique de l'aurore boréale* (Paris, 1733, 1754), interpreted them instead as the luminous matter of the solar atmosphere fallen toward the earth. The "electric" hypothesis to which Darwin alludes was notably advanced by Benjamin Franklin (e.g., *Exper. Elec.*), John Canton (*Phil. Trans.* 48 [1753–54]: 350–58) and Priestley (see *LP* 1.373n) (*The History and Present State of Electricity* [London, 1767]).

1.410 Dryads Wood-nymphs.

1.429–30 Dove ... Tremella See *LP* 1.393.

1.437 laurel *Prunus*; see *LP* 3.40.

1.438 myrtle *Myrtus*, "held sacred to the goddess Venus and ... regarded as a symbol of love, peace, honor, etc." (*OED*); see *LP* 1.57.

1.441 thorn Usually, the hawthorn.

1.444 clue Or clew, a ball of thread.

1.446 fretted Ridged, as the fingerboard of a lute or other stringed instrument.

Interlude [1]

Interlude 1.2 flower-painter ... portraits of history Academic art in the eighteenth century—that which bore the imprimatur of such institutions as the the *Académie royale de peinture et de sculpture* in Paris (est. 1648) and the Royal Academy in London (est. 1768)—conformed to a hierarchy of generic prestige. Epic or historical painting was considered the most poetical, epitomizing not only stylistic excellence but civic virtue. Portraiture ranked lower, still life ("flower painting") and landscape lowest of all. The subjects of history-painting might be either ancient or modern, though the latter were often portrayed allegorically. Accomplished practitioners of the art in Britain included Benjamin West (American, 1738–1820) and James Barry (Irish, 1741–1806).

Interlude 1.5 sublimity The sublime was an aesthetic mode that attracted considerable interest during the eighteenth century. The difference between the sublime and a related category, the beautiful, was generally understood as the difference between awe-inspiring wildness and tranquility, between irregularity and smoothness, between the ability to inspire thoughts of death and the ability to inspire thoughts of love. Dating back at least to Longinus (*On the Sublime*, c. first century CE), the concept was theorized by Edmund Burke (Irish, 1729–97) (*A Philosophical Enquiry into the Origin of Our Ideas of the Sublime and the Beautiful*, 1757) and, later, Immanuel Kant (German, 1724–1804) (*Critique of Judgment*, 1790). In art it was most closely associated with the landscapes of Salvator Rosa (Italian, 1615–73); in literature, with Shakespeare, Milton and, eventually, the gothic novelist Ann Radcliffe (English, 1764–1823).

Interlude 1.6 Warwick ... would not fly *3 Henry VI* 5.2.32–33, slightly misquoted.

Interlude 1.8–9 Mr. Pope ... Eels *renown'd* Pope, "Windsor-Forest" (London, 1713), 341, slightly misquoted.

Interlude 1.14 Mr. Gibbon's history ... with extensive forests Only the abridged, two-volume edition of Edward Gibbon's (English, 1737–94) *History of the Decline and Fall of the Roman Empire* (London, 1789) speaks of the "immense forests which overshadowed part of Germany and Poland" (1.127).

Interlude 1.14 Mr. Burk's eloquent orations Burke (see Interlude 1.5) served as Member of Parliament for Bristol (1774–80) and for Malton (1780–94). A long-time Whig, he was sympathetic to the American revolutionaries and the movement toward Irish independence; he was also instrumental in the impeachment of Warren Hastings, governor-general of Bengal, for maladministration in imperial India. His speeches and other writings were circulated widely, even before his scathingly anti-Jacobin *Reflections on the Revolution in France* (1790) brought him his greatest notoriety.

Interlude 1.16 "The Lady herself ... part of her" Perhaps Ovid, *Remedia amoris*: "In all the specious figure which you see,/The least, least part of her own self is she" (trans. Nahum Tate, 370–71). Darwin would repeat the sentiment in his *Plan for the Conduct of Female Education* (1797), sect. 28.

Interlude 1.16 Mason's English Garden William Mason (English, 1725–97), *The English Garden*, 4 vols. (London, 1772–81); based on Virgil's *Georgics* (37–30 BCE).

Interlude 1.16 Hayley's Epistles William Hayley (English, 1745–1820), poet, biographer and patron of the arts; his verse-epistles include *Epistle on Painting* (1778), *An Essay on History* (1780) and *An Essay on Epic Poetry* (1781). He also provided a dedicatory poem to *3EV1795+* (see textual notes).

Interlude 1.21–23 "She never told ... her damask cheek" *Twelfth Night* 2.4.110–12.

Interlude 1.25–32 "To Reason ... my name" This song, with slightly different lyrics, first appears in *The Merry Man's Companion* (London, 1750); other compilations give the performer as Thomas Lowe (English, c. 1719–83) or John Beard (1716/17–91). The words may have been written by John West (English, 1693–1766), later first Earl De La Warr and Viscount Cantelupe.

Interlude 1.33 Rubens ... Luxemburgh gallery Marie de Médici, widow of Henry IV, commissioned Peter Paul Rubens (Flemish, 1577–1640) to adorn her recently constructed Palais du Luxembourg in Paris. Rubens painted twenty-four canvases (1622–25) commemorating Marie's regency through a mixture of portraiture and allegory. The Palais would function as the first public art gallery in France, opening its doors on a limited basis from 1750 to 1779, with Rubens' cycle as the chief attraction. It paved the way for the establishment of the Musée du Louvre in 1793, and would itself be established as the Musée du Luxembourg in 1818.

Interlude 1.34 Angelica Kauffman Celebrated artist (Swiss, 1741–1807), known especially as a portraitist and history-painter; one of the thirty-six founding members, only two of them women, of the Royal Academy.

Interlude 1.34 Roubiliac Louis-Francois Roubiliac (French, 1702–62), sculptor; his monument to Major-General James Fleming (1683?–1751) stands in Westminster Abbey.

Interlude 1.35 Sir Joshua Reynolds ... Mr. Fielding ... Mr. Garrick Reynolds (English, 1723–92), painter, art theorist and first president of the Royal Academy; in the latter capacity, he delivered fifteen addresses between 1769 and 1790 that were published, both individually and collectively, as *Discourses*. In the *Discourse* to which Darwin refers (London, 1787), Reynolds says: "The lower kind of Comedy, or Farce, like the inferior style of Painting, the more naturally it is represented, the better; but the higher, appears to me to aim no more at imitation, so far as it belongs to any thing like deception; or have any expectation that the spectators should think the events there represented, are really passing before them" (18–19). He goes on to fault Henry Fielding's (English, 1707–54) comic novel *Tom Jones* (1749) for a scene (bk. 16, ch. 5) in which the schoolmaster, Mr. Partridge, is moved to a passion by a performance of *Hamlet* starring the great David Garrick (English, 1717–79).

Interlude 1.36–39 dreams The ancient practice of oneirocriticism, or interpretation of dreams as prophetic, survived into the eighteenth century. In general, however, physiological explanations became more prevalent. Some writers even characterized dreaming as a kind of pathology whose absence constituted "perfect" sleep. Nightmares were conceptualized as having not only physical origins (luxurious diet; lying on one's back; retention of fluids, especially in women) but potentially dangerous physical consequences; notable commentators included George Cheyne (*The English Malady*, 1733), John Bond (*Essay on the Incubus*, 1753) and William Buchan (*Domestic Medicine*, 1769). Such causes, however, did not have to be proximate: according to Locke (*Essay Concerning Human Understanding*, 1689), dreams consisted of sensory data accumulated by the waking mind, but assembled in so haphazard fashion as to suspend rational selfhood; according to David Hartley (*Observations on Man*, 1749), the sensory stimuli could be remembered rather than immediate. One more supernaturally inclined outlier was Andrew Baxter (*Enquiry into the Nature of the Human Soul*, 1733), who claimed that the sleeping body, being closed off from the mind, was left vulnerable to impositions by spirits; dreams are perceptions, therefore, but these perceptions are not one's own. There was widespread agreement that what was true of the dreaming mind could also be true of the waking, as in cases of madness, religious zealotry, political "enthusiasm" or, as Darwin says here, artistic imagination. See, crucially, *LP* 3.51–78 and n; as well as *Zoonomia*, Vol. 1, Chapter 18, "Of Sleep." In the latter, Darwin also elaborates on the idea that sleep makes us more susceptible to internal stimuli but less susceptible to external ones; indeed, the latter change produces the former, by allowing the replenishment and "accumulation of sensorial power." Darwin's meditations on the science of sleep influenced such early psychobiologists as Pierre-Jean-Georges Cabanis (*Rapports du physique et du moral de l'homme*, 1802) and Franz Joseph Gall (*Anatomie et physiologie du système nerveux*, 1810–19). Coleridge also pays them unacknowledged homage in an 1818 lecture on Shakespeare's *The Tempest* ("sleep ... consists in a suspension of the voluntary, and therefore of the comparative, power"), although in general he regards Darwin as too much the materialist.

Interlude 1.41 Elements of Criticism by Lord Kaimes Henry Home, Lord Kames (Scottish, 1696–1782), jurist and philosopher; *Elements of Criticism* (Edinburgh, 1762), vol. 1, ch. 2 ("Emotions and Passions"). Kames defines "ideal presence" as "*a waking dream*" that "vanisheth upon the first reflection of our present situation" (108). Such a "reverie" is more "faint" than "real presence", which, being "vouched by eye-sight, commands our belief, not only during the direct perception, but in reflecting afterward upon the object"; yet it is stronger than a "general or reflective remembrance" (108, 112). The person under its influence, "losing consciousness of himself ... conceives every incident as passing in his presence, precisely as if he were an eye-witness"; and experiences "passions" that, unlike simple "emotions", rouse him to action (112). It can be conjured by painting, books or, especially, theater.

Interlude 1.45 Fusseli Henry Fuseli, born Johann Heinrich Füssli (Swiss, 1741–1825), artist and writer who served as professor of painting at the Royal Academy. He contributed four designs to the *Botanic Garden*: *Flora Attired by the Elements* (Illustration EV1), *The Fertilization of Egypt* (Illustration EV4), *Tornado* (Illustration EV4a; *3EV1795*) and *The Nightmare* (Illustration LP8a; *5LP1799*). See also *LP* 3.56.

Interlude 1.45 "his ear must ... on Lethe's brink" "I find thee apt;/And duller shouldst thou be than the fat weed/That roots itself in ease on Lethe wharf,/ Wouldst thou not stir in this" (Ghost, *Hamlet* 1.5.31–34).

Interlude 1.46 machinery "Contrivances employed for effect in a literary work", especially "supernatural personages and incidents" (*OED*); as the spirits attendant on Prospero in Shakespeare's *The Tempest* (1611), or the Rosicrucian beings in Darwin's own poem.

Interlude 1.47 Mr. Brydone's travels Patrick Brydone (Scottish, 1736–1818), *A Tour through Sicily and Malta*, 2 vols. (London, 1773). Letter 22 describes the "amazing crowd of statues" assembled by the prince Francesco Ferdinando Gravina e Alliata at Villa Palagonia, Bagheria, Sicily: the prince "devoted his whole life to the study of monsters and chimeras", and "put the heads of men to the bodies of every sort of animal, and the heads of every other animal to the bodies of men" (2.54).

Canto II

2.5 Orb within orb Ancient astronomers, notably Eudoxus of Cnidus (fourth century BCE), had conceptualized the universe in terms of concentric spheres to which the planets were attached. This model was long in force, although opinions differed as to whether it corresponded to a material reality or was merely descriptive of the planetary paths. Johannes Kepler (German, 1571–1630) was instrumental in replacing a theory of orbs (*orbes*) with a theory of orbits (*orbitae*). By Darwin's time, references to the former are clearly poetical.

2.11 Iris Goddess of the rainbow; attendant on Juno, the wife of Jupiter.

2.15 north Presumably, the location of Polaris, the pole star, in the constellation Ursa Minor.

2.18 "Bear on broad ... to the skies" Duplicates *LP* 2.8.

2.19–24 Plume after plume ... upon the winds Inexact metaphor, but the import seems to be that Carlina (the *fair Mechanic*, *LP* 2.20) has furnished wings (*pennons*, *LP* 2.23) for her seeds, just as Daedalus (*LP* 2.20) did for himself and his son Icarus from feathers (*eider* or duck *down*, *LP* 2.21), thread (*silken strings*, *LP* 2.21) and wax (the *Whale-bone ribs*, *LP* 2.20, are Darwin's addition). Daedalus had built a labyrinth for King Minos of Crete and was later imprisoned there with Icarus; the wings enabled them to escape. Icarus flew too close to the sun, melted the wax and drowned; Daedalus landed safely in Sicily.

2.25–66 So on the ... it steer'd, directs The brothers Joseph-Michel (1740–1810) and Jacques-Étienne Montgolfier (French, 1745–1799) performed the first public launch of an untethered hot-air balloon on 4 June 1783 at the Annonay marketplace, inflating their device with smoke from a straw-fed fire. They repeated the demonstration on 19 September at Versailles with animal passengers, and on 21 November over Paris with the passengers Jean-François Pilâtre de Rozier and François Laurent, Marquis d'Arlandes. The balloon used in the first flight was made of sackcloth and lined with paper (of which the brothers' father, Pierre Montgolfier, was a respected manufacturer); those in the subsequent flights, varnished taffeta. "Balloonomania" ensued, encompassing a host of further inventions and experiments (starting in August 1783, the physicist Jacques-Alexander-César Charles made his own trials on a hydrogen-filled balloon made of rubberized silk); speculations as to how the technology might be applied in times of both peace and war; literary commentaries both panegyric and satiric; and fashions sartorial and tonsorial. Joseph-Michel was the only Montgolfier brother to make a balloon flight himself, which he did from Lyons on 10 January 1784.

2.47 the pearly star Venus.

2.54 the Georgian star See *LP* 1.189.

2.57 headlong Comets Isaac Newton (English,1642/43–1727), in his *Philosophiae Naturalis Principia Mathematica* (1687), demonstrated that the "attractive"

(*LP* 2.55) force of the Sun had caused the great (i.e., observable) comet of 1680 to move along a conic section. Halley (see *LP* 1.403), in his *Synopsis of the Astronomy of Comets* (1705), studied the documentation of twenty-four comets and determined that their orbits were ellipses rather than open parabolas. He thus predicted, accurately, that a comet last observed in 1682 would reappear every seventy-five or -six years; it bears his name today.

2.59 Cassiope Cassiopeia, northern constellation identified with the enthroned Ethiopian queen of Greek mythology.

2.60 the Bear Ursa Major, northern constellation identified by the Greeks with the ursiform nymph Callisto; the group consisting of its seven brightest stars is known popularly as the Plough, the Big Dipper or Charles' Wain.

2.60n Tibi jam brachia ... Virg. Georg. "Already ardent Scorpio contracts its claws for you"; Virgil, *Georgics*; addressed to Caesar.

2.63 Argo Argo Navis, southern constellation identified with the ship *Argus* in which Jason sought the golden fleece (see *LP* 1.349–58).

2.67 Lina See *LP* 1.65n.

2.67n Curcuma See *LP* 1.65.

2.67n Isis Egyptian goddess of the Earth; sister/wife of Osiris, god of the Nile.

2.67n Arachne Greek maiden who challenged the goddess Athena (Roman Minerva) to a weaving contest. Athena chose to depict the triumphs of the gods; Arachne, their failings. Offended, Athena destroyed Arachne's work. Arachne then hanged herself in shame, but Athena took pity on her and turned her into a spider. See Ovid, *Metamorphoses* 6.1–145.

2.67–71 Woof ... weft ... warp See *LP* 1.346.

2.68 shuttle In weaving, a boat-shaped device used for throwing the threads of the woof or weft across those of the warp. The fly-shuttle, invented by John Kay in 1733, enabled the production of textiles whose width exceeded a single weaver's arm-span.

2.70 Quick beat the reeds The *reed* of a loom is a row of vertical separators (originally made of reed) between which the threads of the warp are passed; it is held by parallel bars that, when moved along the threads of the warp, *beat* those of the weft into place.

2.70 pedals Treadles; used to raise and lower different sets of warp-threads, allowing for the passage of the shuttle and for the creation of woven patterns.

2.71 beam Roller around which the excess warp is wound. Absent one, *massy weights* (*LP* 2.72) can be used to keep the warp taut.

2.74–75 Isis ... Arachne See 2.67n.

2.77–80 *Five* sister-nymphs ... the dizzy wheel Flax may be hand-spun by simply rolling the raw fiber into a strand with moistened fingers; or by holding the raw fiber on a stick or distaff, drawing it out, twisting it into a strand, then winding it around a spindle (a drop-spindle hangs freely and puts an additional twist in the thread as it spins). The sixteenth-century "Saxon" spinning wheel, instead of a distaff, uses a tall stationary rod; and, instead of a spindle, uses a bobbin rotated by a great wheel, which in turn is operated by a foot-treadle. In 1764, James

Hargreaves invented the spinning jenny, which winds the thread on eight or more separate spindles. The thread may then be woven on a loom.

2.78 beamy "Massive as a (weaver's) beam" (*OED*); raw flax forms a stiff bundle.

2.85 Derwent River in Derbyshire; not to be confused with the River Derwent in Cumbria, later commemorated in William Wordsworth's *Prelude* (1799, 1805, 1850).

2.87n Sir Richard Arkwright English, 1732–92; the mills at Cromford, near Matlock, Derbyshire, were built on land that he and his partners leased in 1771. He patented his water-powered spinning machine, which also used the *rollers* Darwin describes here, in 1769; and his carding engine, for combing out long fibers, in 1775. Arkwright's machines yielded a relatively strong cotton yarn suitable for warp, whereas the relatively weak yarn from Hargreaves' spinning-jenny was suitable only for weft. During the 1780s, Arkwright was embroiled in legal disputes over his patents, with Darwin appearing as a witness on his behalf.

2.87n Hemp *Cannabis*.

2.87n Mulberry-tree *Morus*.

2.87n parenchymatous substance Pulp or pith.

2.105 Papyra *Cyperus papyrus*.

2.105n invention of music, of astronomical observation Sometimes attributed, respectively, to Jubal (Genesis 4.21) and to Enoch (1 Enoch 72–82, apocryphal).

2.105n Whence arose the policy ... profession of his father See Montesquieu, *The Spirit of Laws* (1748; London, 1750), vol. 2, book 20, ch. 20: "Laws, which oblige every one to continue in his profession, and to devolve it to his children, neither are nor can be of use in any but despotic kingdoms; where no body either can, or ought to have emulation" (17). Also Rousseau, *Emilius* (London, 1762), vol. 1, book 1: "In Egypt, where the son was obliged to follow the occupation or profession of his father, education had at least a determinate end; but among us, where rank and profession are only permanent, and persons continually changing; a father would not know whether, in bringing up his child to his own profession, he might be doing him good or ill" (13).

2.105n cyphers Numerals (see *LP* 2.126).

2.105n crotchets Possibly, "musical characters"; technically, quarter-note symbols (see *LP* 2.130).

2.105n emphasis, and tone, and swell The use of dynamic markings in European music only started becoming widespread in the mid-eighteenth century.

2.105n Charles the twelfth ... numeration by squares Cf. Voltaire, *The History of Charles XII, King of Sweden* (1727; Edinburgh, 1769), book 8: "He wanted to alter the method of counting by tens, and proposed to substitute in its place the number sixty-four, because that number contains both a square and a cube, and being divided by two is reducible to a unit. This, if it proves any thing, only shews that he always delighted in what was difficult and extraordinary" (340). In 1716, Charles (1682–1718) enlisted the help of the philosopher Emanuel Swedenborg

(1688–1772), who suggested a simpler base-eight system; the digits 0 through 7 would be assigned alphabetic characters, which could be strung together into "words" to express larger numbers. Before his scheme could be implemented, Charles was shot dead at the invasion of Norway.

2.105n to reform our alphabet Semantic, grammatical and orthographic inconsistencies within and across the languages of Europe had been a matter of longstanding concern. In eighteenth-century Britain, one could find descriptivist and prescriptivist commentators; prescriptivists who sought to standardize language and radically to reform it; reformers who dreamt of a "universal language" that would eliminate semantic differences and of a "universal character" that would eliminate orthographic ones; universalists who accepted a language of arbitrary signs and who aspired to a "real" or "natural" language; natural-language advocates who recommended a phonetic alphabet and a pictorial one. Prominent phoneticians included Benjamin Franklin ("Scheme for a New Alphabet and a Reformed Mode of Spelling", c. 1768) and Thomas Spence (*The Grand Repository of the English Language*, 1775). In Darwin's immediate circle, Priestley also pondered these issues (*A Course of Lectures on the Theory of Language, and Universal Grammar*, 1762).

2.123 surd Unvoiced (literally, "deaf") consonant.

2.127 decimating In military parlance, to "decimate" is to "select by lot and put to death one in every ten of (a body of soldiers guilty of mutiny or other crime): a practice in the ancient Roman army" (*OED*).

2.130 quaver Musical eighth-note symbol.

2.133 cane See *Canna*, *LP* 1.39.

2.138 clarion "A shrill-sounding trumpet with a narrow tube" (*OED*).

2.145 Hermetic From Hermes Trismegistus, "the name given by the Neo-platonists and the devotees of mysticism and alchemy to the Egyptian god Thoth, regarded as more or less identified with the Grecian Hermes [see *LP* 2.287], and as the author of all mysterious doctrines, and especially of the secrets of alchemy" (*OED*).

2.148 chalybeate Iron-infused; from the Chalybes, ancient people of Asia Minor known as ironworkers.

2.153 Delany Mary Granville Pendarves Delany (English, 1700–88), artist. She assembled her exquisite botanical collages mostly during the eighteen years (1768–86) that she lived with Margaret Harley, Duchess of Portland, on the latter's estate at Bulstrode, Buckinghamshire. These "paper-mosaicks" were bequeathed to the British Museum at the end of the nineteenth century; 985 of them reside there today.

2.153n Hortus siccus "Dry garden."

2.153n Mrs. North Probably Henrietta Maria Bannister North (1750–96), wife of Brownlow North, Bishop of Winchester. She is credited as the namesake of *Moraea* (also *Marica*) *northiana* in Henry Andrews' *The Botanist's Repository*, vol. 4 (London, 1797), which adds that "to [her] fervor and liberality in botanical pursuits we owe much of the present prevailing taste for the science" (pl. 255).

2.163 Nymphæa See *LP* 1.177n.

2.163n Horologe Clock; section 335 of Linnaeus' *Philosophia botanica* explains the *Horologium florae*.

2.163n Stillingfleet's calendar of Flora See *LP* 1.288n. Stillingfleet's *Miscellaneous Tracts*, 2nd ed. (London, 1762) includes three articles titled "The Calendar of Flora": one "Swedish and English, made in the year 1755" (translated from Alexander Malachias Berger's *Calendarium florae, Amoen. acad.* 67 (31 March 1756): 387–415); one "made at Stratton in Norfolk" (by Stillingfleet himself, 1755); and one "by Theophrastus, at Athens" (translated from *Peri phyton historia*, c. 300 BCE).

2.166 nutation Here, "the periodic variation of the inclination of the earth's axis" (*OED*).

2.171–76 First in its ... ballance pants below The power-source of a mechanical watch is a steel mainspring. As it uncoils, it transfers energy to the train, or system of gears to which the hands are attached. This in turn transfers energy to the escapement, a toothed wheel that allows the watch to advance in stepwise fashion. The balance, an oscillating wheel that constitutes the timekeeping element, both regulates the motion of the escapement and is kept in motion by it. Portable watches were first manufactured in Germany around 1500; brass mechanisms superseded iron around 1580; the balance-springs introduced by Robert Hooke (see *LP* Proem 2) and Christiaan Huygens (Dutch, 1629–95) greatly improved accuracy; the lever escapement invented by Thomas Mudge (English, 1717–94) was detached from the balance, allowing it to oscillate more freely.

2.199n nectaries See *LP* 1.69n.

2.199n Tropaeolum See *LP* 4.45.

2.209–26 So from his ... his pendant ears Nebuchadnezzar, King of Babylonia (Daniel 1–4), refused to recognize any power greater than his own; he was stripped of his kingdom, "and he was driven from men, and did eat grass as oxen, and his body was wet with the dew of heaven, till his hairs were grown like eagles' feathers, and his nails like birds' claws" (4.33). Eventually he came to his senses, praised God and was restored.

2.219 brinded Brindled, dark-streaked or -spotted.

2.227n Saint Anthony ... travels in Italy Joseph Addison's (English, 1672–1719) *Remarks on Several Parts of Italy, &c. in the Years 1701, 1702, 1703* (London, 1705) includes a translated quotation from a biography of St. Anthony sold at Padua, recounting his sermon to the fishes: "He had no sooner done speaking, but behold a miracle! The Fish, as tho' they had been endu'd with Reason, bow'd down their Heads with all the Marks of a profound Humility and Devotion, moving their Bodies up and down with a kind of Fondness, as approving what had been spoken by the Blessed Father, St. *Antonio*. The Legend adds, that after many Hereticks who were present at the Miracle had been converted by it, the Saint gave his Benediction to the Fish and dismissed 'em" (74).

2.229 rice *Oryza*.

2.238 fell See *LP* 1.131.

2.238 to taste is death Darwin himself was an almost-lifelong teetotaler, and wrote to James Hutton in 1778 that "Vinous spirit from small beer to alcohol destroys us all" (*Letters* 78–7). See also *LP* 3.287–300 and n.

2.251 Torpedoes Electric rays.

2.253 Phocae Seals.

2.254 Grampi Dolphins or other small ceteceans.

2.266 asphodel *Asphodelus*.

2.266 amaranth *Amaranthus*.

2.268 Papaver See *LP* 1.69n. In eighteenth-century Britain, the medicinal and recreational use of opium was legal and virtually unrestricted. Tremendous profits were to be made, moreover, from the black-market sale of India-grown opium to China, which had officially banned the drug in 1729 in response to widespread addiction. The cultivation in Bengal was supervised by the East India Company itself; shipping, by freelance merchants; distribution, by local Chinese smugglers. Opium imports totaled roughly 1,300 chests per year by 1780 and succeeded in reversing the historic British trade deficit with China.

2.268n The emaciated and ... Baron de Tott François, Baron de Tott (French, 1733–93), *Memoirs...Containing the State of the Turkish Empire and the Crimea, during the Late War with Russia* (London, 1785), describes "a part of Constantinople, called Teriaky Tcharchissy, or the Market for the takers of Opium"; these "pale and melancholy Figures would be sufficient to raise our pity, did not their lengthened Necks, their Heads turned on one side, their Back Bone distorted, their Shoulder raised up to their Ear, and a number of other extravagant attitudes which result from their Disease, exhibit a Picture of the most ridiculous Nature" (1.141–42).

2.280 Parian Of Paros, Greek island in the Cyclades, known for its white marble.

2.287–90 So with his ... Pluto's dreary reign *Hermes* (Roman Mercury), son and messenger of Zeus, invented the lyre from a tortoise-shell and gave it to Apollo; Apollo in turn gave him the *caduceus*, a winged staff with two serpents coiled around it. Zeus appointed him usher of souls to the underworld of *Pluto* or Hades.

2.291 Crewe Emma Crewe (English, c. 1768–1850), amateur artist and designer of pottery for Josiah Wedgwood (English, 1730–95).

2.301n *Cistus lubdaniferus* Presumably *Cistus ladaniferus*.

2.301n Cactus grandiflora ... Cerea See *LP* 4.15.

2.301n Hibiscus trionum See *LP* 1.373n.

2.301n Hibiscus sinensis, mutable rose Perhaps *Hibiscus rosa-sinensis* (China rose) or *Hibiscus mutabilis* (mutable); Darwin's description seems to fit the latter.

2.321 Severn River flowing from eastern Wales to the Bristol Channel.

2.323 Pan Greek fertility-god, represented as a piper with the upper body of a man and the lower of a goat.

2.323 Eudness Apparently a wood near Bridgnorth, Shropshire, in the Severn Valley.

2.324n gamut Full range of notes in a musical system, specifically that of Guido d'Arezzo (c. 990–1050), which involved seven overlapping hexachords (six-note scales); contraction of *gamma ut* (the lowest recognized note in this system).

2.330 rime See *LP* 1.254.

2.343 Andes Mountain system extending along the entire western coast of South America.

2.345n Several of these ... this famous drug Loja (Loxa), in the south of present-day Ecuador, was founded in the mid-sixteenth century, destroyed by an earthquake in the seventeenth and rebuilt on a new location. The Quechua Indians there knew the medicinal properties of the bark they called *kina-kina* (whence later *quinine*, its active ingredient). Ana, Countess of Chinchon and the Spanish vice-queen of Peru, introduced the bark to Europe after being thus cured of malaria at Lima in 1638; the Jesuit priests furthered its dissemination (whence its popular name *Jesuits' powder*); the physician Thomas Sydenham (c. 1624–89) popularized its use in England; Linnaeus named it *Cinchona* in the (misspelled) Countess' honor. The story told by Darwin appears in G. Douglas, *A Treatise of the Fossil, Vegetable and Animal Substances That Are Made Use of in Physick* (London, 1736), 303–04; based on Étienne-François Geoffroy, *Tractatus de materia medica* (1699).

2.347 Quito Capital of present-day Ecuador; about 230 miles north of Loxa, as mentioned by several eighteenth-century tracts on Peruvian bark.

2.353 Loxa See *LP* 2.345n.

2.355 Hygiea Also Hygeia, Greek goddess of health.

2.358 The star of Autumn Probably Sirius (see *LP* 1.327); Pope's *Iliad* describes "Orion's dog (the year when autumn weighs)", whose "burning breath/ Taints the red air with fevers, plagues and death" (22.39, 41–42).

2.359 Ague "High fever ... marked by an intense feeling of cold and shivering" (*OED*).

2.369 Hygiea See *LP* 2.355.

2.392 ocherous Ochreous; containing or of the (usually yellow-brown) color of ochre, an iron ore.

2.401–10 Thus Israel's ... the almighty Power See Exodus 17.1–6.

2.411 Bolster'd Propped up, as by a bolster or cushion.

2.412 Dropsy Edema, a "morbid condition characterized by the accumulation of watery fluid in ... the serous cavities [i.e., lined with fluid-secreting membranes] or the connective tissue of the body" (*OED*) and sometimes associated with excessive thirst.

2.415 Tantalus Son of Zeus and Pluto, he was admitted to the banquet of the gods and sinned against them (in various versions, by revealing their secrets to mortals, by stealing their food and delivering it to mortals, and by serving them his butchered son Pelops to test their knowledge). His eternal punishment was to stand up to his neck in water that retreated whenever he tried to drink, with overhanging fruit that the wind blew away whenever he tried to eat.

2.421n ascites Abdominal dropsy.

2.421n rectified spirit of wine "A mixture of ethanol and water purified by repeated distillation" (*OED*).

2.421n Experiments on mucilaginous ... by Cadel, London Charles Darwin (English, 1758–78; Erasmus' son), *Experiments Establishing a Criterion between Mucilaginous and Purulent Matter; and an Account of the Retrograde Motions of the Absorbent Vessels of Animals Bodies in Some Diseases*, ed. Erasmus Darwin (London: T. *Caddell*, 1780).

2.421n mucilaginous Slimy, viscous.

2.421n purulent Pus-like.

2.425 serpent-wreathed wand Caduceus, a physician's symbol; see *LP* 2.287–90.

2.429 mephitic Foul-smelling, noxious.

2.431 and n Marseille's good Bishop ... escaped the disease A bubonic plague in 1720–22 claimed approximately half the population of *Marseille* (city and seaport in southeastern France) and its surroundings, or 80,000 lives. Henri-François-Xavier de Belsunce de Castelmoron (French, 1671–1755), a Jesuit, became known as the "*Good Bishop*" for his ministrations during the epidemic. Darwin closely reproduces language from "Marseilles", John Barrow, *A New Geographical Dictionary*, 2 vols. (London: J. *Coote*, 1759–60).

2.431 and n London's generous Mayor ... *R. Exchange*. 1722. The outbreak referred to as the Great Plague killed over 75,000 people, an estimated 20 per cent of the total population of London, between 1664 and 1666. The bills of mortality for 1665 alone report 68,596 deaths. The cause of plague has been identified as the infectious bacterium *Yersinia pestis*, transmitted from rats to humans through flea bites. Sir *John Lawrence* (d. 1692), Lord Mayor of London in 1664–65, was widely praised for his management of the crisis. Notable plague accounts include the diary of Samuel Pepys (English, 1633–1703); and Daniel Defoe (English, 1660?–1731), *A* Journal of the Plague-Year (London: "Printed for E. *Nutt at the Royal Exchange*; J. Roberts in Warwick-Lane; A. Dodd without Temple-Bar; and J. Graves in St. James's Street", 1722). Defoe's novel claims to be a memoir written shortly after the plague by a survivor, one "H.F.", and presents itself through detailed first-person observations and accounts of public records and documents. See also *LP* 3.317–42.

2.436 Zembla Novaya Zemyla, archipelago in northwestern Russia.

2.438 northern lustres See *LP* 1.403.

2.442 Howard John Howard (English, 1726?–90), philanthropist. Moved by his experience as high sheriff of Bedfordshire, he became a crusader for prison reform, visiting hundreds of institutions throughout Britain and the continent over nearly two decades. Of particular concern to him was the prevalence of physical and moral disease in the prisons, and the practice of re-incarcerating acquitted prisoners who could not afford their discharge fees. He facilitated the passage of Parliamentary acts providing for the payment of prison fees by the county (1774), the proper hygiene and medical care of prisoners (1774) and penitentiary houses for prisoner rehabilitation (1779), although these measures were inconsistently enforced. Late in life he turned his attention to lazaretti (plague-hospitals for the indigent) and died in the Ukraine of a fever he contracted while inspecting a military hospital.

2.473–80 And now her ... quintessence of Tea Tea was a luxury item when it entered the European market in the seventeenth century. It was first imported from China by the Dutch East India Company (VOC) in 1610, and by the English East India Company in 1669. Its effects on the bodily constitution, the sexual/domestic order and the national economy were praised by some (Nahum Tate, Joseph Addison and Richard Steele) and denounced by others (Daniel Defoe, Jonas Hanway, Eliza Haywood). Darwin's stylized epithets also recall the coffee ceremony in Pope's *Rape of the Lock* 3.105–24.

Interlude [2]

Interlude 2.1 bulls with brazen feet The Colchian king Aetes agreed to yield the golden fleece if Jason would yoke two brazen-footed, fire-breathing bulls and sow the teeth of a dragon slain by Cadmus. Assisted by the sorcery of Medea, he succeeded. See *LP* 1.349–58.

Interlude 2.1 fire-breathing dragons ... the Hesperian fruit The earth-goddess Gaia presented Zeus and Hera with a wedding gift of golden apples. Hera placed the fruit in a paradisiacal island garden to be guarded by the nymphs called the Hesperides (daughters of Hesperus) and the dragon Ladon. The eleventh labor of Hercules was to retrieve the apples. Aphrodite also gave three of them to Hippomenes, who had challenged the virgin huntress Atalanta to a footrace; he won it, and her hand, when she bent down to pick the apples up.

Interlude 2.2 prints of the London Cries *The Cryes of the City of London Drawn after the Life* (1687), a popular collection of forty comical prints depicting London street-vendors "crying" or hawking their wares. Drawn by Marcellus Laroon, engraved by John Savage and published by Pierce Tempest, it includes such types as "Sow-Gelder", "Mountabanck" and "The merry Milk Maid." The practice itself was much discussed in the eighteenth century as a public nuisance, notably in Addison's *Spectator* 251 (Tuesday 18 December 1711).

Interlude 2.3 quadrate i.e., be "square", agree. In her *Memoirs of the Life of Dr. Darwin* (London: J. Johnson, 1804), Anna Seward (English, 1742–1809) remarks of this passage: "if Homer's similies [sic] do not often touch the object with which they are compared, at all points, yet they are never so utterly without connexion with it, as several which may be found in this poem. That a poetic simile ... is the more sublime, or more beautiful, for not quadrating exactly; yet it ought to possess such a degree of affinity with the subject, that ... we may feel, though we cannot verbally demonstrate the perfect justness of the similitude" (328–29).

Interlude 2.5 a philosophical analogy ... instead of poetry See Darwin's "Advertisement" (moved from *LP* for *1EV1792+*) which, somewhat differently, describes an interplay between these modes.

Interlude 2.5 picturesque Used with increasing frequency to describe an aesthetic mode distinct from the sublime and the beautiful (see *LP* Interlude 1.5), as involving a harmonious irregularity. It was notably theorized in the travel-writings of William Gilpin (English, 1724–1804) such as *Observations...on Several Parts of England* (London, 1786) and *Observations on the River Wye* (London, 1789), Uvdale Price's (English, 1747–1829) *Essay on the Picturesque* (London, 1794) and Richard Payne Knight's (English, 1750–1824) *The Landscape, A Didactic Poem* (London, 1794); and informed the landscape architecture of Humphry Repton (English, 1752–1818), with its penchant for cultivated wildness.

Interlude 2.7 Certainly, he will ... the night-mare See *LP* Interlude 1.36–39 and 3.51–78.

Interlude 2.8 the cruelty of ... flaying Marcias alive Marsyas took up a flute thrown away by its inventor, Minerva; discovering that he had a great talent for music, he challenged Apollo to a competition. Apollo, on the lyre, was victorious; he bound Marsyas to a tree and skinned him for his impudence. Modern depictions include those by Titian (Italian, c. 1488–1576), Bartolomeo Manfredi (Italian, c. 1582–1622), Jusepe de Ribera (Spanish, 1591–1652) and Luca Giordano (Italian, 1634–1705).

Interlude 2.11 has not this ... safe on land? Lucretius, *De rerum natura* 2.1–4: "'Tis pleasant, safely to behold from shoar/The rowling Ship; and hear the Tempest roar:/Not that another's pain is our delight;/But Pains unfelt produce the pleasing Sight" (trans. John Dryden). The opening phrase of this passage, *Suave mari magno*, became proverbial for serene detachment from a scene of distress.

Interlude 2.11 Akinside ... Pleasures of Imagination Mark Akenside (English, 1721–70), *The Pleasures of Imagination* (London, 1744), 2.155 ff. In a note, Akenside says that the "very mysterious kind of pleasure which is often found in the exercise of passions generally considered painful" is attributable not to Lucretian "self-love" but to "the general delight which the mind takes in it's own activity ... join'd with the moral applause of it's own temper, which attends those emotions when natural and just" (66–67).

Interlude 2.14 The descriptions of ... cave of Polypheme The best-known account is in Homer's *Odyssey* 9; a more lurid one, in Euripides' satyr-play *Cyclops* 375–419.

Interlude 2.14 Scaliger Julius Caesar Scaliger (Italian, 1484–1558), natural historian, grammarian and literary critic; author of *Poetices* (1561).

Interlude 2.14 Titus Andronicus The attribution of this bloody tragedy was widely disputed in eighteenth-century England. Thomas Capell (1713–81) and Thomas Tyrwhitt (1730–86) were in the minority that accepted Shakespeare's authorship. Edward Ravenscroft (1659–97), whose own adaptation *Titus Andronicus, or The Rape of Lavinia* was first produced in 1678; Lewis Theobald (1688–1744); and Thomas Percy (1729–1811) believed that *Titus* was substantially the work of another playwright but improved by Shakespeare. Samuel Johnson (1709–84), who nevertheless included *Titus* in his eight-volume Shakespeare edition of 1765; and his collaborator George Steevens (1736–1800) denied Shakespeare's involvement absolutely.

Canto III

3.1 shell Lyre, legendarily made from a tortoise's shell.

3.7n Sleaford-church St. Denys, built starting in the twelfth century.

3.7n Druids Members of an ancient Celtic priesthood, represented as magicians in Welsh and Irish folklore.

3.7n Bryony *Bryonia*; charlatans sometimes carved, re-buried and sold the fleshy root of its *alba* (white) species as that of mandrake.

3.7n Mandrake *Mandragora*, a genus appearing in Linnaeus but not in *SV*. That the forked mandrake root was said to resemble the human shape may explain its reputation as an aphrodisiac or fertility treatment. Like many of the plants in this canto, it is imagined as either therapeutic or highly poisonous, depending on the dose. Early examples of the legends to which Darwin refers here include Genesis 30 and Josephus' *Wars of the Jews* 7.6.

3.7n Peony *Paeonia*; see *LP* 1.69n.

3.7n Anodyne necklaces Galen (Greek, 129–c. 199 CE) first records this folk-remedy in *De simplicium medicamentorum facultatibus* 6.3, but its dental application was popularized by Paul Chamberlen's (English, 1635–1717) *Philosophical Essay upon Actions on Distant Subjects* (London, 1715).

3.7n Price's History of Cornwall William Pryce (English, c. 1735–90), *Mineralogia Cornubiensis* (London, 1778); Book 3, ch. 1 deals with divining or dowsing.

3.7n hazle *Corylus*.

3.7n Animal Magnatism Term coined by the physician Franz Anton Mesmer (German, 1734–1815) for the attractive force allegedly governing an invisible bodily fluid. Disease resulted from obstructions to this magnetic flow, which could be therapeutically manipulated in a kind of hypnosis that came to be called "mesmerism." An investigative commission convened by Louis XVI in 1784 (Benjamin Franklin and Antoine Lavoisier were members) debunked Mesmer's system, which nevertheless continued to enthrall the imagination of nineteenth-century authors from Charles Dickens to Elizabeth Gaskell to Bram Stoker.

3.8 numbers Metrical verse; music.

3.19 yew *Taxus*; evergreen traditionally planted in churchyards and associated with sorrow.

3.29 choir Or chancel; the portion of a church, between the transept and the altar, where the services are performed.

3.34 mitre ... cope Ecclesiastical headdress and cloak, respectively.

3.40 Pythian Of Delphi (formerly called Pytho); oracular.

3.40 Laura *Prunus*; see *LP* 1.437. The reference to a "wreath" (*LP* 3.44), however, suggests that Darwin may conflate *Prunus laurocerasus* (cherry laurel) with the more familiar *Laurus nobilis* (bay; see *LP* 3.184n), worn as a crown to signify military, athletic or poetic achievement.

3.40n Æn. L. vi In Book (*L[ibrum]*) 6 of Virgil's *Aeneid*, Aeneas visits Deiphobe, the Cumaean Sybil (not the Delphic oracle), who prophesies war and conducts him to the underworld. The poem does contain references to laurel, but not in this context.

3.40n Ratafie Ratafia; a cordial usually flavored with cherry-, peach- or almond-pits, which do contain small amounts of cyanide.

3.40n cherry stones ... Apricot-kernels From *Prunus cerasus* and *Prunus armeniaca*, respectively.

3.40n Peach-leaves From *Amygdalus persica*, now classed as *Prunus persica*.

3.40n Wallnut-leaves From *Juglans*.

3.46 cinctured Girdled.

3.46 ephods Sleeveless priestly vestments.

3.51 Nightmare Incubus; a demon (usually male) believed to press down on the body of a sleeping person (usually female) for purposes of sexual congress. "Mare" as a female horse has a separate etymology, which Darwin apparently misunderstands (perhaps through an erroneous reading of the illustration; see *LP* 3.56) in describing the "Fiend" as "flit[ting]" "on his Nightmare."

3.52 squab Squat.

3.52 fen Marsh.

3.53 wilder'd Lost.

3.56 Fusseli's See Interlude 1.45 and Illustration LP8a.

3.57 Shakespear's happiest grace From 1789 to 1805 the London alderman John Boydell (English, 1720–1804) mounted an exhibition of thirty-four (later, over a hundred) paintings of scenes from Shakespeare, featuring eighteen of the most eminent artists of contemporary Europe. The Shakespeare Gallery proved an enormous success, although Boydell's efforts to publish an accompanying edition of engravings fell flat. Fuseli was responsible for nine works, including two scenes from *A Midsummer Night's Dream*. Darwin alludes (*LP* 3.58) to Theseus' famous description, in that play, of the poet who "gives to airy nothing/A local habitation and a name" (5.1.16–17).

3.74 and n The Will presides not See *LP* Interlude 1.36–39. Since a nightmare is one of the few sleep-events in which the will does become activated, the logic connecting *LP* 3.74 to 3.51–73 is somewhat unclear; perhaps the "bower" in which the "will presides not" is specifically that of what *Zoonomia* calls "perfect sleep." Darwin's concept of "will", however, is inconsistent at best. In *Zoonomia* (e.g. vol. 1, ch. 34, "Diseases of Volition"), he defines it much more categorically as "the active state of the sensorial faculty in producing motion in consequence of desire or aversion: whether we have the power of restraining that action, or not"; by this light, "voluntary" motions would seem to characterize "perfect" and "imperfect sleep" alike.

3.74n tube-rose *Polianthes tuberosa*.

3.77 Gorgon In Greek mythology, a monstrous female with hair of snakes and a petrifying gaze; Medusa was one of three.

3.80n egged i.e., ovate; broad at the base and tapering at the end.

3.80n Sloan. Hist. of Jamaica *Voyage*, vol. 2, book 1, ch 7.

3.80n Lin. Spec. Plant. See *LP* 1.131n; class 12, section 16.

3.80n Capri-ficus See *LP* 4.328.

3.85 Hamps and Manifold Except during the flood season, these two rivers in the Derbyshire Peak District disappear underground for part of their length, the Hamps from Waterhouses and the Manifold from Wetton, until they converge with the River Dove near Ilam and Thorpe (see *LP* 3.123–30). The earlier name of the Hamps, *Hanespe*, derives from a Celtic phrase meaning "dry in summer."

3.89 Nymphs See *LP* 1.1.

3.90 Thor Thor's house, a cave located inside a limestone crag in the Derbyshire Peak District, may have been occupied as early as the Upper Paleolithic age. Darwin's dramatic account in *LP* 3.101–08 notwithstanding, its name most likely derives not from the Norse god of thunder but from *tor* (hill) or *thurse* (goblin); *thurse-house* and *thurse-hole* are topographical provincialisms.

3.90n Wetton ... Hamps and Manifold ... Ilam See *LP* 3.85.

3.90n Jasper See *AN* 19.23.

3.90n Limestone See *EV* 2.35n.

3.90n Calcareous Spars See *EV* 2.93n.

3.90n Agate See *EV* 2.215n.

3.90n corallines Algae (or compound animal organisms erroneously believed to be plants, as were corals) whose shoots contain calcareous deposits.

3.90n Marble See *EV* 2.93n.

3.90n Lead See *EV* 2.398n.

3.90n Copper See *EV* 1.279n.

3.90n Zinc See *EV* 2.403.

3.90n Flint See *EV* 1.229n.

3.90n Chert See *EV* 2.299.

3.90n Toadstone See *AN* 6.2.

3.90n Lava See *EV* 1.139n.

3.90n Whitehurst's ... Hutton's Theories of the Earth John Whitehurst (English, 1713–88), geologist and original Lunar Society member; his *Inquiry into the Original State and Formation of the Earth* (London, 1778) is notable chiefly for its appendix on Derbyshire strata, which—like James Hutton's (Scottish, 1726–97) "Theory of the Earth" (*Edinb. Trans.* 1 [1788]: 209–304)—advances largely Plutonist and uniformitarian views (see *AN* 20.1, 20.4). Curiously, however, Darwin seizes upon the catastrophist elements of Whitehurst's tract here, misattributing the Derbyshire caves (which were in fact formed by the action of water on soluble limestone) to subterranean explosions. The District did see active volcanoes during the Carboniferous period (c. 300 million years ago), but their effect was primarily to convert the organic limestone into mineral form.

3.92 dome The Peak District has this overall shape, which becomes Darwin's point of departure for an architectural conceit: *piles* (*LP* 3.93), columns; *piers* (3.95), pillars; *ribs* (3.96), arches, as in Gothic vaults.

3.102 fane Temple.

3.119–22 sparry ... Lava ... coral ... jasper ... agate See *LP* 3.90n.

3.124 impending Overhanging.

3.131 Impatiens See *LP* 1.194n.

3.131n beard Or *awn*, bristle in which the grain-sheath of a grass terminates.

3.131n hygrometer Or hygroscope, device for measuring atmospheric humidity; a rudimentary version was invented by Leonardo da Vinci (Italian, 1452–1519), and an improved one using human hair by Horace-Bénédict de Saussure (Swiss, 1740–1799). An experiment involving the wild-oat beard appears in Hooke's *Micrographia* (London, 1665), Observation 27.

3.131n barley *Hordeum*.

3.131n Mr. Edgeworth ... wooden automaton Darwin describes the first of two hygrometers sent to him by the inventor and educational theorist Richard Lovell Edgeworth (English, 1744–1817). Locomotive machines appear to have fascinated Edgeworth: among his other inventions contemplated or realized were one-wheeled, four-wheeled, spring-wheeled, wind-operated and steam-operated conveyances; as well as an eight-legged wooden horse that might vault its rider over a wall. His daughter, the novelist Maria Edgeworth (English, 1768–1849), discusses contemporary hygrometry debates at length in her tale "Harry and Lucy" (incorporated into her collection *Early Lessons* in 1803).

3.135–78 So when Medæa ... with convulsive laugh After Medea helped *Jason* restore the golden fleece to his native *Thessaly* (see *LP* 1.349–58), they resided at Corinth as husband and wife. He betrayed her with *Creusa*, the daughter of King Creon. Medea took her revenge by murdering Creusa, Creon, and her own children by Jason; setting fire to the palace; then fleeing for Athens in a serpent-drawn chariot. The best-known account is by Euripides (431 BCE), who gives Medea two sons rather than a son and a daughter.

3.159 horrent Bristling.

3.160 furies Goddesses of vengeance.

3.184n volatility Tendency to evaporate at ordinary temperatures.

3.184n Tobacco *Nicotiana*.

3.184n cloves *Caryophyllus*; see *LP* 4.207.

3.184n cinnamon *Laurus cinnamomum*, now classed as *Cinnamomum verum*.

3.184n balsam of peru Resin from *Myroxylon peruiferum*.

3.184n tar Dark viscous liquid produced by the distillation of an organic substance such as wood or coal.

3.184n pitch Dark resin produced by the distillation of tar.

3.184n turpentine See *LP* 1.73n.

3.184n M. de Thosse Not identified, but his paper is "Mémoire sur la manière de détruire les pucerons qui attaquent les arbres frutiers", *Mémoires d'agriculture, d'économie rurale et domestique* (Spring 1787): 106–11.

3.184n pucins ou pucerons Aphids.

3.185 weird Trafficking in fate.

3.188n deadly nightshade *Atropa belladonna*.

3.188n henbane *Hyoscyamus*.

3.188n houndstongue *Cynoglossum*; see *LP* 1.194n.

3.188n Ilex See *LP* 1.143.

3.188n Dictamnus See *LP* 3.184.

3.188n science of medicine ... principal materials in *Zoonomia*, Part III, Article II, Darwin classes the cherry laurel, tobacco, deadly nightshade and henbane among the "Incitantia", or substances "which increase the exertions of all the irritative motions."

3.189 hebenon generically, poison; possibly from *henbane* (see 3.188n) or *ebon* (black). The next line shows that Darwin is thinking of the ghost in *Hamlet*: "Upon my secure hour thy uncle stole/With juice of cursed hebona [*Q1*: Hebenon *F1*] in a vial/And in the porches of my ears did pour/The leprous distillment" (1.5.61–64).

3.191 mad'ning throng cf. John Dryden, trans. *Aeneid* (1697): "She flies the town, and mixing with the throng/Of madding matrons, bears the bride along" (7.538–39); Thomas Gray, "Elegy Written in a Country Church-Yard" (1751): "Far from the madding crowd's ignoble strife" (73); Jonathan Swift, "Ode on Science" (1763): "Behold the madding throng,/In wonder and oblivion drown'd" (45–46).

3.191 Urtica See *LP* 1.73n. When the nettle pierces the skin it injects histamine ($C_5H_9N_3$), which causes itching; and serotonin ($C_{10}H_{12}N_2O$) and acetylcholine ($C_7H_{16}NO_2$), which cause a burning sensation.

3.191n Hook. microgr. See *LP* 3.131n; Observation 44, "Of the tufted or brush-horn'd Gnat."

3.191n Marsi and Psilli (Psylli) Ancient tribes of central Italy and North Africa, respectively, known among the Romans as snake-charmers.

3.191n Beccaria Possibly Father Giambattista Beccaria (Italian, 1716–81), Professor of Physics at the University of Turin, author of *Dell' Elettricismo Naturale ed Artificiale* (1753); elected Fellow of the Royal Society in 1755, he was instrumental in disseminating the scientific thought of Benjamin Franklin throughout Europe. May also refer to Jacopo Bartolomeo Beccari (Italian, 1682–1766), Bolognese physician and natural philosopher; see *EV* 1.182 and n.

3.193n Ingenhouz exper. on air Jan Ingenhousz (Dutch, 1730–99), *Experiments upon Vegetables, Discovering Their Great Power of Purifying the Common Air* (London, 1779), which first propounded the theory of photosynthesis.

3.193n Jacquini hort. botanic. Vindeb. Nikolaus Joseph Jacquin (Dutch, 1727–1817), *Hortus botanicus vindobonensis* [The Viennese Botanic Garden] (Vienna, 1770–76), 1.10.

3.193n Boerhaave ... Mead ... Grotto del Cani The "grotto" anecdote appears both in Herman Boerhaave's (Dutch, 1668–1738) *Institutiones et Experimenta Chemiae* (Paris, 1724), 1.214; and in Richard Mead's (see *LP* 1.61n) *A Mechanical Account of Poisons in Several Essays* (London, 1702), 153. *Grotta del Cane* (Cave of the Dog) is located at the city of Pozzuoli (as is the volcanic crater Solfatara; see *AN* 20); carbon dioxide issues from crevices in its floor. In a popular demonstration for tourists, a dog would be placed inside and allowed to approach the point of suffocation.

3.193n the gas from fermenting liquors ... limestone the conversion of sugar into ethyl alcohol (C_2H_5OH) releases CO_2, as does the heating of limestone ($CaCo_3$).

3.197 Palmyra Tadmor, an ancient city in central Syria; autonomous for roughly two millennia, it came under Roman control in the first century CE. A *decumanus* or east-west street, lined with Corinthian columns, leads from the great temple of Baal Shamen.

3.197n Volney's travels Constantin-François de Chasseboeuf, comte de Volney (French, 1757–1820), *Voyage en Syrie et en Égypte* (Paris, 1787), ch. 30.

3.205 rill Small brook.

3.225 nutmeg *Myristica.*

3.226 plaintain If "towering", presumably *Musa* (plantain-tree; see *LP* 1.313n) rather than *Plantago* (plantain; see *LP* 1.77).

3.228 chaplet Wreath worn on the head.

3.229 lichen Symbiotic combination of algae and fungi.

3.233 refluent Flowing back.

3.234 revolant Flying back, a Latinism; Darwin's line is the sole *OED* quotation.

3.236 irremeable Allowing no return.

3.238 and n Upas Not in Linnaeus; now classed as *Antiaris toxicara*. Its latex is used to poison arrows. Although its existence had been known to Europeans since at least the fourteenth century, Darwin was at the cutting edge of rumors concerning its more fantastical properties. The *London Magazine* published a "Description of the Poison-Tree, in the Island of Java", credited to "N.P./J.N. [Johannes Nicolaus] *Foersch*, Translated from the original Dutch, by Mr. [Carl] Heydinger" (n.s. 1 [December 1783]: 511–17). Foersch, a surgeon in the Dutch East India Company (VOC), seems to have discussed the matter previously with Royal Society President Sir Joseph Banks, per a letter dated 24 July ?1776: "I assure you that what I had the honour to tell you about the upac [sic] tree incredible as it seems to be to you, has ben [sic] no more then [sic] plain and real truth...." The upas proved a potent metaphor thereafter: see, e.g., Mary Wollstonecraft, *A Vindication of the Rights of Woman*, 1792, Chapter 8 ("those vices, which like the Java tree shed a pestiferous vapour around—death is in the shade!"; 309); William Blake, "A Poison Tree", *Songs of Innocence and of Experience*, 1794; Sir Francis Burdett, speech of 25 January 1810 (a corrupt Parliament as the "root of all the evil [whose] branches spread over and extended to every extremity of the country. Under their shade flourished no useful plants, nothing but noxious weeds"; *Edinburgh Annual Register* 3.1.22); Percy Bysshe Shelley, *Queen Mab*, 1813 (4.83–89, 4.260–65, 5.44–52 and 6.207–08); and Samuel Taylor Coleridge, *Notebooks*, 11 November 1827 (corruption as spreading its "Upas arms over the whole Garden of the Church"; 5.5662) and December 1829 (sodomy as "this hideous Upas-tree of the Moral World"; 5.6160). See also *LP* Additional Notes.

3.238 Hydra Monstrous serpent in Greek mythology; if any of its many heads were cut off, two would grow in its place. Its blood and breath were highly poisonous. Hercules destroyed it as one of his labors, cutting off each head in turn and having his servant, Iolaus, burn the stump.

3.242 leagues See *LP* 1.232n.

3.248 pounce Puncture.

3.259n cotyledons See *LP* 1.373n.

3.259n Reichard Johann Jacob Reichard (German, 1743–82), botanist on whose 1778 Frankfurt edition of Linnaeus' *Genera Plantarum* (1737) Darwin relied for his 1787 translation, titled *Families of Plants*. The references to *Species Plantarum* appear to be erroneous.

3.259n The Rev. T. Martyn ... letters on botany Thomas Martyn (English, 1735–1825), professor of botany at Cambridge University and member of the Royal Society; he translated and continued Jean-Jacques Rousseau's (Swiss, 1712–78) *Lettres éléméntaires sur la botanique* (Geneva, 1781–83) as *Letters on the Elements of Botany* (1785).

3.259n incommode Bother, disturb, or inconvenience.

3.259n Ivy *Hedera.*

3.259n Helxine Linnaeus' term for several species of *Polygonum* (see *LP* 1.86n); now applied to *Soleirolia.*

3.259n Convolvulus See *LP* 2.163n.

3.259n Cynanche Unrecognized form of *Cynanchum*; Darwin the physician may be thinking of the throat disease by this name.

3.259n Asparagus The edible variety is *Asparagus officinalis.*

3.265 Scamander Greek river god, personified in Homer's *Iliad*, book 20 as a river that flows from Mount Ida through Troy; called Xanthos by the gods and Skamandros by men. Scamander fights on the side of the Trojans, and attempts to kill Achilles, the hero of the Greeks, after Achilles insults him.

3.265 Laocoon A Trojan priest of Posiedon (Neptune), the god of the sea. He defied Poseidon's rules by marrying and having sons. In Virgil's *Aeneid*, Laocoön warns the Trojans against accepting the Trojan horse with the famous line "Timeo Danaos et dona ferentes" ("I fear Greeks, even those bearing gifts") and hurls his spear into the side of the horse, which emits a telltale hollow sound (*Aeneid* II.13–56). This angers the goddess Minerva, who sends sea-serpents to strangle Laocoön and his sons (*Aeneid* II.199–233). Laocoön's death is the subject of a famous first-century marble sculpture, *Laocoön and his Sons* (or the "Laocoön Group"), attributed by Pliny the Elder to Athanadoros, Hagesandros and Polydoros, all of Rhodes. This sculpture occasioned the German dramatist and critic Gotthold Ephraim Lessing's (1729–81) *Laokoön, oder über die Grenzen der Malerei und Poesie* (1766)—a significant contribution to modern discussions of aesthetics, including Darwin's (see *LP* Interlude 3.13).

3.287 Vitis The common grape vine is *Vitis vinifera*. See also *LP* 2.238.

3.287n mucilage A viscid starch.

3.287n Mahomet The seventh-century prophet Muhammed, founder of Islam. Among the advances made by Muslim chemists and alchemists (eminent ones included Jabir ibn Hayyan, or Geber, c. 721–815; and Abu Bakr Muhammad ibn

Zakariya al-Razi, or Rhazes, c. 865–925) was the development of processes and apparatus for distillation. One of the substances they are credited with discovering is ethanol—pure alcohol or drinking alcohol—though Islam forbids its consumption. The words *alchemy*, *alcohol* and *alembic* (a distillation vessel) are all of Arabic origin.

3.287n theory of ... Cadell While a medical student, Darwin's son Charles (see *LP* 2.421n) won first prize from the Aesculapian Society of Edinburgh in 1778 for his thesis on pus and mucus; after Charles died suddenly later that year, Erasmus arranged for it to be published. *Experiments Establishing a Criterion between Mucaginous and Purulent Matter, and an Account of the Retrograde Motions of the Absorbant Vessels of Animal Bodies in Some Diseases* was published in 1780 at Edinburgh (for William Creech), Lichfield (for John Jackson), and London (for Thomas *Cadell*). Erasmus "edited" the octavo volume; supplied a "life of the author"; and most likely wrote the "Retrograde Motions" section (among others), which deals with alcohol-induced "*Diabœtes and Dropsy*."

3.290 Thyrsus Staff carried by Dionysus (Bacchus) and his followers, usually a stalk of giant fennel (*Ferela communis*) covered with ivy and topped with a pine cone. Dionysus is the god of grapes, wine, fertility and theater; his frenzied festivals are dedicated to freeing the self from care. In botany, a thyrsus (thyrse) refers to a compact type of panicle, or branched flower cluster.

3.294 mantling Gathering a scum or coating, like a mantle (cloak).

3.295 Chemia Ancient Greek word meaning "transmutation" and usually referring to what would later be called alchemy; here, perhaps, the genius of chemistry.

3.297 Gout a metabolic disease that presents itself in recurring attacks of inflammatory arthritis; caused by elevated levels in the blood of uric acid, which becomes crystallized in the joint areas and can spread to other internal organs. Also known as podagra (Greek, "foot-trap") because of its common initial appearance in the big toe. Darwin attributes it to "intemperance in drinking", rather than to "gluttony in eating" as is commonly thought (*Zoonomia* vol. 1, ch. 21, "Of Drunkenness"); present-day explanations are primarily genetic.

3.298 Dropsy the accumulation of fluid beneath the skin or in the cavities of the body; now termed edema (Greek, "swelling"). See *Zoonomia* vol. 1, ch. 28, "Of the Paralysis of the Absorbent System"; and vol. 1, ch. 29, "Of the Retrograde Motions of the Absorbent System", which links dropsy with diabetes.

3.299 Lepra Greek, "scaly"; leprosy, an infectious and chronic disease that affects body tissues. Presenting itself initially as skin lesions, it is often followed by decreased circulation and sensation, especially in the extremities. Permanent damage to the skin, peripheral nerves, mucous membranes and muscles can result. Since the cause and transmission of leprosy were long unknown, sufferers experienced significant social stigma and sometimes segregation into leper camps. In 1873 the physician G.H. Armauer Hansen (Norwegian, 1841–1912) pinpointed the bacillus *Mycobacterium leprae* as the causative agent of leprosy, which has henceforth been known as Hansen's Disease. *Mycobacterium lepromatosis*, a distinct species, was identified in 2008. See *Zoonomia* vol. ch. 26, "Of the Capillary Glands and Membranes."

3.300 Frenzy see *Zoonomia* vol. 1, ch. 33, "Diseases of Sensation", on the sources and types of "delirium."

3.301–10 and n So when Prometheus ... remorseless fangs Prometheus was one of the titans in Greek mythology. He is best known for stealing the secret of fire from Zeus (*Jove*), the king of the Gods and lightning, and giving it to humans. As punishment, he was bound to a rock in the *Caucasus*, a region of mountains at the border of Europe and Asia, by the blacksmith-god Hephaestus (*Vulcan*). Zeus sent an eagle every day to eat out Prometheus' *liver*, which would regenerate each time, creating a cycle of eternal torment. In *Zoonomia*, Darwin reaffirms that the myth may have been invented by ancient physicians as an allegory for the effects of excess alcohol consumption on the liver (vol. 1, ch. 30, "Paralysis of the Liver and Kidneys"); Shelley would endorse a similar view in his long note to *Queen Mab* 8.211–12 (1813), published separately as "A Vindication of Natural Diet." Darwin's note also hints at his theory of evolution through the inheritance of acquired characteristics; see *Zoonomia* vol. 1, ch. 39, "Of Generation."

3.301n calcining Burning in an attempt to purify.

3.301n scirrhous Tumorous.

3.311 and n Cyclamen See *LP* 1.61n. Known as *sow-bread* because swine feed on its fleshy roots; Darwin confuses it with *shew-bread*, a religious offering.

3.314 inhumes Buries.

3.317 Plague See *LP* 2.431 and n.

3.324 Cleone Possibly inspired by the titular heroine of a popular tragedy (1758) by Robert Dodsley (English, 1704–64); she too loses a child, although the similarities between the two stories end there.

3.338 and n Chartreuse *La Grande Chartreuse*, a Carthusian monastery founded in London in 1371; its name was corrupted to "Charterhouse." Today, the building is a charitable institution known as Sutton's Hospital in Charterhouse. From 1348 to 1371 the site had served as a mass grave for victims of the Black Death, a plague that devastated Europe in the fourteenth century, with the most deaths between 1348 and 1350. The passage in Defoe's *Journal* to which Darwin refers, however, describes a mass burial not at the Charterhouse but at a churchyard near Whitechapel (see *LP* 2.431 and n).

3.343 Ontario The smallest and easternmost of the five freshwater Great Lakes of North America.

3.345n mimosa scandens See *LP* 1.267; this "climbing" species is now classed as *Entada phaseoloides*. Its seeds are known as cacoons (*cocoons*).

3.345n cocoa-nuts *Cocos nucifera*. One of the *Palmae*, listed in the Appendix to *SV* (i.e., not among Linnaeus' twenty-four classes).

3.345n Dr. Tonning Henrik Tonning (Norwegian, 1732–96), *Rariora Norvegiae* (27 February 1768), *Amoen. Acad.* 7.149 (1769): 466–86.

3.345n Gulf-Stream Clockwise oceanic current that flows from the Gulf of Mexico around the southern tip of Florida, up the east coast of the United States and past Newfoundland; some of its eddies then continue toward the British Isles and the Norwegian seas, where they form the North Atlantic Drift. It was named in 1762 by Benjamin *Franklin* (American, 1706–90), who, while serving in England as deputy postmaster general for the colonies, heard complaints that the British mail took too much time crossing the Atlantic. Timothy *Folger* (American, 1732–1814), Franklin's cousin and the captain of a Nantucket merchant vessel, sketched the currents of the Gulf Stream based on data from ship logs; Franklin then published the map around 1769–70.

3.345n Dr. Blagden Sir Charles Blagden (English, 1748–1820), "On the Heat of the Water in the Gulf-stream", *Phil. Trans.* 71 (1781): 334–44.

3.345n Governor Pownal Thomas Pownall (English, 1722–1805), governor of the Massachusetts Bay Colony from 1757–59 and later M.P. for Minehead, Somerset; a map in his *Hydraulic and Nautical Observations on the Currents in the Atlantic Ocean* (London, 1787) shows the Gulf Stream, based on his communication with Franklin.

3.345n Sir Hans Sloane See *LP* 1.313n; "An Account of Four Sorts of Strange Beans, Frequently Cast on Shoar on the Orkney Isles", *Phil. Trans.* 19 (1695–97): 298–300; "Strange Beans Frequently Cast Ashore on the Orkney's", *Phil. Trans....Abridged* 3 (1721): 540–41.

3.345n Phaseolus See *LP* 3.259n; but Sloane, who calls these seeds "cocoons", may be describing what Linnaeus would classify as *Mimosa scandens* (see above).

3.345n Horse-eye bean Most likely Linnaeus' *Dolichos pruriens*, now classed as *Mucuna pruriens*.

3.345n Niker Nicker, nickar; most likely Linnaeus' *Guilandina bonduc*, now classed as *Caesalpinia bonduc*.

3.345n Lenticula marina, or Sargosso Most likely Linnaeus' *Fucus natans* (see *LP* 4.161), now classed as *Sargassum natans*.

3.351–64 So the sad ... of the Nile See Exodus 1.22–2.10. The Pharaoh of Egypt (capital, *Memphis*) commands that all newborn Hebrew males be drowned. When she can no longer keep him hidden, Jochebed (*the sad mother*) sends her son Moses (the *dear babe*) floating down the *Nile* in a basket woven from bulrushes or papyrus reeds (*Paper-flags*). His sister, Miriam, watches from a distance until the basket reaches a spot where the Pharaoh's daughter is bathing. Moses then grows up in the royal family as brother to the future Pharaoh.

3.365–70 Erewhile majestic ... rent the sky See Exodus 2.11–4.20. The adult Moses (*the Prophet*) flees to Midian (*his lone abode*) after killing an Egyptian taskmaster. He sojourns there incognito, marries and works as a shepherd until the Lord appoints him, reluctant, to lead the Israelites out of slavery.

3.374 Fierce SLAVERY stalks Although slavery had been effectively ruled illegal in England with the James Somersett case of 1772, the slave trade continued apace. In 1787 was formed the Society for the Purpose of Effecting the Abolition of the Slave Trade, a predominantly Quaker organization emphasizing education, Christian principle and legislative action. Among the influential texts it distributed was co-founder Thomas Clarkson's (English, 1760–1846)

Summary View of the Slave Trade and of the Probable Consequences of Its Abolition (London, 1787) and the freed slave Olaudah Equiano's (Nigerian? c.1745–97) *Interesting Narrative* (London, 1789). Though Darwin was not one of the twelve founding members of the Society, he supported its efforts. His friend, the industrialist Josiah Wedgwood, produced cameos featuring the seal for the Society, which depicted a supplicating black man in chains with the text "Am I Not a Man and a Brother?" (see *EV* 2.297–320, *EV* 2.421–30 and Illustration EV3).

3.374 and slips the dogs of hell cf. Shakespeare, *Julius Caesar*: "And Caesar's spirit, ranging for revenge ... Shall in these confines with a monarch's voice/Cry 'Havoc!' and let slip the dogs of war" (3.1.270–73).

3.377 Ye bands of Senators In 1789, the Yorkshire M.P. and prominent abolitionist William Wilberforce (English, 1759–1833) introduced twelve resolutions against the slave trade in what was reported to be one of the most moving speeches given in the House of Commons. In 1791, the motion he brought to the House to abolish the slave trade was defeated 163–88, but through continuing efforts on behalf of Wilberforce and his abolitionist supporters, a bill to abolish the slave trade in the British West Indies passed in 1807.

3.378 either Ind I.e., the West Indes and the Indian subcontinent. British colonial activity in both began in the early seventeenth century; India achieved independence in 1947.

3.383 still small voice After sending wind, an earthquake and fire, the Lord speaks to the prophet Elijah at Mount Horeb in "a still small voice" (1 Kings 19.12).

3.388 "He, who allows oppression, shares the crime" Frequently and dubiously attributed to the theologian and humanist Desiderius Erasmus (Dutch, 1466–1536). Something like this sentiment appears in Erasmus' *Copia*, but it is a second-hand reference to Cicero's *Pro Cluentio*. Cicero argues that Oppianicus, who bribes his brother-in-law's widow to procure an abortion, deserves harsher punishment than the woman herself.

Interlude [3]

Interlude 3.1 sister-art Horace's *Ars poetica* is the source of the famous formulation *ut pictura poesis* (361–65), and the kinship between the disciplines was enunciated even earlier in Aristotle's *Poetics*. The metaphor of poetry as the "sister" to painting appeared in Charles-Alphonse du Fresnoy's (French, 1611–68) *De arte graphica* (1668), which John Dryden (English, 1631–1700) translated with an original preface as *The Art of Painting* (London, 1695). Dryden had previously used the phrase in "To the Pious Memory of the Accomplisht Young Lady Mrs. Anne Killigrew, Excellent in the Two Sister-Arts of Poësie, and Painting" (1685) and in "To Sir Godfrey Kneller" (1694).

Interlude 3.5–9 When we have ... upon the land *Midsummer* 2.1.128–32. "Following (her womb then rich" is more conventional than "Following her womb, (then rich." Darwin's reading is suggested, albeit punctuated differently, in the Samuel Johnson–George Steevens edition of 1778.

Interlude 3.10 At the end ... described in words *Lear* 5.3.283, with "Pray you" for "Pray, Sir." Nahum Tate (English, c. 1652–1715) made a notorious adaptation of the play in 1681, giving it a happy ending.

Interlude 3.12 The ingenious Bishop ... wrinkles on it George Berkeley (Irish, 1685–1752), Bishop of Cloyne, *An Essay towards a New Theory of Vision* (Dublin, 1709). In this early work, Berkeley adumbrates the "heterogeneity thesis" that the ideas arising from the different senses are of fundamentally different kinds. The objects of vision operate as a "language" in that they arbitrarily signify the objects of touch but have no essential connection with them.

Interlude 3.13 But in one ... representing stationary ones The idea of poetry as a temporal and painting as a spatial medium was most famously articulated in Lessing's *Laokoön*; see *LP* 3.265.

Interlude 3.14 climacteric Building to a climax.

Interlude 3.14 In the representation ... of Ant. Coypel Antoine Coypel (French, 1661–1722), *Sacrifice of Jephthah*; Darwin may have seen an engraving by Gaspard Duchange (French, 1662–1757). To fulfill a vow, Jephthah the Gileadite made a burnt offering of his daughter after his victory over the Ammonites (Judges 11.30–40).

Interlude 3.16 gamut See *LP* 2.324n.

Interlude 3.16 recitativo Style of sung speech used primarily to advance the narrative of an opera or oratorio.

Interlude 3.16 chaunting The singing "in public worship" of "the Psalms, Canticles, etc." to a "short melody or phrase" (*OED*).

Interlude 3.16 stops Consonants "in the formation of which the passage of the breath is completely obstructed" (*OED*).

Interlude 3.16 sibilant "Having a hissing sound" (*OED*).

Interlude 3.16 If the Chinese ... are acquainted with The tonal properties of the Chinese language had been noted by westerners at least since Matteo Ricci (Italian, 1552–1610), the influential Jesuit missionary.

Interlude 3.17 There is however ... measure or time The discussion that follows is notationally idiosyncratic, though not unprecedented; see, e.g., John Mason, *An Essay on the Power and Harmony of Prosaic Numbers* (London, 1749). Darwin's first and second examples use poetic meters conventionally regarded as duple (iambic); his third and fourth, as triple (dactylic). To translate the former into a triple musical meter and the latter into a duple, he increases the duration of the stressed syllables, evidently aspiring to a melody or, perhaps, to a classical quantitative meter rather than a modern accentual one (see *LP* Interlude 3.35). That Darwin may allow rhetorical phrasing to predominate over a strict meter is also suggested by his awkward barring in the first and second examples, which creates a syncopated rhythm by routinely assigning the downbeat to an unaccented syllable (instead of "allowing a note or to preceed the commencement of the bar"; *LP* Interlude 3.35). Finally, Darwin sets his third, fourth and fifth examples in 4/4, not 2/4, although the term "common time" was used for both.

Interlude 3.18–19 Soft-warbling beaks ... the inchanted grove See *LP* 4.221–22.

Interlude 3.21–22 Life buds or ... kindles, as it rolls Unidentified; presumably Darwin's invention.

Interlude 3.24 Mr. Anstie's Bath-guide Christopher Anstey (English, 1724–1805), *The New Bath Guide; or, Memoirs of the B—r—d Family, in a Series of Poetical Epistles* (London, 1766). Darwin misquotes from Letter 11 ("A Description of the Ball"): "But as near as I guess from the size of his calf,/He may weigh about twenty-three stone and a half"; "His pigeon-wing'd head was not drest quite so soon,/For it took up a barber the whole afternoon." The character in question is named "Marmozet."

Interlude 3.30 Shenstone William Shenstone (English, 1714–63), "The Rape of the Trap" (1737), describing a rat.

Interlude 3.37 Sir Isaac Newton ... the following numbers *Opticks* (London, 1704), Proposition 6, Problem 2. Newton had advanced this analogy as early as 1675, in a letter to the Royal Society; Hooke had noted it even earlier, as had the polymath Athanasius Kircher (German, 1601–80; see *LP* 1.101). The music theory here is complicated. The solfege syllables (Sol, La, etc.) indicate that Newton is using an eight-degree scale in the Dorian mode (modern D-D) but ascending it by means of six-degree hexachords (see *LP* 2.324n), which overlap but do not repeat. To complete a progression such as Newton describes here, one must mutate (jump) among hexachords, resulting in the duplication of certain syllables. As for Newton's numerical fractions, they reflect the tuning system known as just intonation, albeit incorrectly for a D-D scale (they would be correct for a C-C or a G-G scale). Assume a monochord (single string) with a length of 1; as one ascends the scale, the intervals between successive tones can be represented by the ratios 8/9, 9/10, 15/16, 8/9, 9/10, 15/16, 8/9. Each ratio represents the fraction of the string length of one tone that is required to generate the next tone. (Modern notation uses Hertz instead of string lengths, and so inverts the ratios.) For example, if one starts with a string that gives C, taking 8/9 of the C length gives D; taking 9/10 of the D length gives E; etc. Thus, the size of each interval can be represented as the fraction by which the string for one tone must be shortened to get to the next tone; whence Newton's progression 1/9, 1/10, 1/16, 1/9, 1/10, 1/16, 1/9. This is a different schema from the one earlier in the *Opticks* (Book 1, Part 2, Proposition 3, Problem 1, Experiment 7), where the ratios represent the interval from each tone to the root tone, not to the neighboring tone.

Interlude 3.39 Dr. Smith Robert Smith (English, 1689–1768), mathematician and natural philosopher; *Harmonics, or The Philosophy of Musical Sounds* (London, 1749), Section 4, Article 7.

Interlude 3.40 Mr. Argand Aimé (Ami) Argand (Swiss, 1750–1803), inventor of a smokeless oil-lamp with a circular wick and glass chimney. He patented it in

England in 1784; members of the Lunar Circle were involved in its manufacture. See also *Letters* 86–3 ff.

Interlude 3.41 Mr. Guyot Edmé-Gilles Guyot (French, 1706–86), "Musique oculaire", *Nouvelles recreations physiques et mathématiques*, vol. 3 (Paris, 1770), no. 52; trans. William Hooper, *Rational Recreations* (London, 1774).

Interlude 3.41 Father Castel Louis-Bertrand Castel (French, 1688–1757), Jesuit priest who explored the prospect of an "ocular harpsichord" (*clavecin pour les yeux*) from the 1720s to the 1750s. Diderot and the composer Jean-Philippe Rameau were supportive; Voltaire and Rousseau were not. Ultimately, Castel seems to have been interested more in the theoretical implications of such an instrument than in overcoming the considerable technical obstacles to its realization.

Interlude 3.42 Dr. Darwin of Shrewsbury Robert Waring Darwin (Erasmus' son, 1766–1848), "New Experiments on the Ocular Spectra of Light and Colours ... Communicated by Erasmus Darwin", *Phil. Trans.* 76 (1786): 313–48.

Interlude 3.42 the blind man ... of a trumpet Locke, *Essay Concerning Human Understanding*, bk. 3, ch. 4 § 11.

Interlude 3.44 Otherwise though the ... than the original E.g., "the word *dentata* applied to a root signifies *teethlike*, applied to leaves it means *toothed*; the word *end-nicked* gives a more accurate idea than the word *emarginatum*, the words *south-turned* than *adversum*, and *one-ranked* than *secunda*" (*SV* x). On *SV*, see *LP* 1.131n.

Interlude 3.46–47 Pale Melancholy sits ... a dread repose Pope, *Eloisa to Abelard* (1717): "Black Melancholy sits, and round her throws/A death-like silence, and a dread repose" (lines 165–66).

Interlude 3.50–51 But round the ... guards the land Seward (see Interlude 2.3), *Elegy on Captain Cook* (London, 1780), p. 13.

Interlude 3.54 It may be ... his cultivated fruit In the earlier eighteenth century, the charge of plagiarism was generally confined to cases of verbatim appropriation, and even then was distinct from the legally sustainable charge of copyright infringement. Other kinds of borrowing could be tolerated, even encouraged, in the spirit of satire and classical learning. In his *Conjectures on Original Composition* (1759), Edward Young (English, 1683–1765) posited imitation of the universal, as opposed to the particular, elements of another's writings as the very definition of originality. From about the 1770s, however, attitudes shifted toward a greater emphasis on originality, a more expansive definition of literary property and thus a heightened sensitivity to plagiarism. Ironically, Darwin himself was not above the practice. In her *Memoirs of the Life of Dr. Darwin*, Seward recalls composing a short poem on Darwin's botanic garden near Lichfield, which he sent "to the Gentleman's magazine [May 1783: 428] and in her name...but, without consulting her, he had substituted for the last six lines, eight of his own. He afterwards, and again without the knowledge of their author, made them the exordium to the first part of his poem [*EV* 1.1–40 *passim*]....No acknowledgment was made that those verses were the work of another pen....They are somewhat altered...and eighteen lines of his own are interwoven with them" (132).

Interlude 3.55 The four lines ... Yonge's Night-thoughts See *LP* 3.255–58; cf. Edward Young, *The Complaint; or, Night-Thoughts on Life, Death & Immortality*, Night the First (London, 1742): "Each *Moment* has its Sickle, emulous/Of *Time*'s enormous Scythe, whose ample Sweep/Strikes Empires from the root; each *Moment* plays/His little weapon in the narrower sphere/Of sweet domestic Comfort, and cuts down/The fairest bloom of sublunary Bliss" (191–96).

Interlude 3.55 The line in ... Justice of Peace See *LP* 3.362; cf. John Langhorne (English, 1735–79), *The Country Justice. A Poem. By One of His Majesty's Justices of the Peace for the County of Somerset*, Part the First (London, 1774): "The big Drops mingling with the Milk He drew" (p. 18).

Canto IV

4.8 Paphian "Of or relating to love and sexual desire; *esp*...illicit" (*OED*); after Paphos, Cyprus, the birthplace of Venus.

4.11 and n Pleas'd Lichfield Darwin's original botanic garden was a love-offering to his future wife Elizabeth Pole; he built it in a valley he had purchased in 1777 one mile west of his Lichfield home. Seward's tribute to the garden (see *LP* Interlude 3.54) was, according to her, what inspired Darwin's own poem.

4.15n Veracrux Veracruz; state, city and port in eastern Mexico.

4.15n Martyn's Letters See *LP* 3.259n.

4.15n Cistus labdaniferus See *LP* 2.301.

4.15n Geranium triste Now classed as *Pelargonium triste*.

4.15n tragopogon See *LP* 2.163n.

4.15n cucubalus See *LP* 1.131n.

4.27 Cynthia's Diana's, the moon's.

4.33–42 Thus, when old ... livelier accents sings See *LP* 1.143n.

4.45n Miss E.C. Linneus Elisabeth Christina (Lisa Stina) Linné (Linnaea) (Swedish, 1743–82), Linnaeus' first daughter; "Om Indianska Krassens Blickande", *Kongl. Vet.-Acad. Handl.* (1762): 284–87.

4.45n Mr. Wilcke Johan Carl Wilcke (German, 1732–96), lecturer in physics at the Swedish Academy; he translated Franklin's *Exper. Elec.* into German as *Des Herrn Benjamin Franklins Esq. Briefe von Elektricität* (Leipzig, 1758).

4.45n Pulteney's view of Linneus Richard Pulteney (English, 1730–1801), *A General View of the Writings of Linnaeus* (London, 1781).

4.53 aerial lights *Ignis fatuus*, will-o'-the-wisp or jack-o'-lantern, a "phosphorescent light seen hovering or flitting over marshy ground, and supposed to be due to the spontaneous combustion of an inflammable gas ... derived from decaying organic matter" (*OED*); known to lead travelers astray.

4.55–72 So when thy ... gem-incircled head King Nebuchadnezzar had Shadrach, Meshach and Abednego cast into a burning fiery furnace, heated seven times greater than usual, for their refusal to worship the golden idol he had erected. He beheld them unharmed and accompanied by a fourth man "like the Son of God" (Daniel 3.25). He ordered that their God henceforth be respected, but his own repentance was not yet at an end; see *LP* 2.209–26.

4.73n *Ovena* *Avena*; see *LP* 1.194n.

4.73n Anthoxanthum See *LP* 1.86.

4.73n wheat See *LP* 1.393n.

4.73n barley See *LP* 3.131n.

4.73n rye *Secale*.

4.73n potatoe of Mexico *Solanum tuberosum* (see *LP* 1.61n), sometimes believed to have originated in Mexico.

4.73n bread fruit *Artocarpus*, native to southeast Asia and the Pacific islands.

4.73n Mr. Wahlborn Johan Gustaf Wahlbom (1724–1808), *Sponsalia plantarum* (1746), *Amoen. acad.* 1 (1749): 61–109.

4.74 Tweed River on the English-Scottish border.

4.95 ozier Osier, *Salix*; see *LP* 1.77n.

4.95 Halcyons See *LP* 1.376.

4.98 Bitterns Small herons.

4.99 beachen Beech, *Fagus*.

4.103n *Galanthus* See *LP* 1.11.

4.103n Stillingfleet's calendar of Flora See *LP* 2.163n.

4.103n salep "A nutritive meal, starch or jelly made from ... dried tubers" (*OED*).

4.103n Gmelin ... history of Siberia Johann Georg Gmelin (1707–55), *Reise durch Sibirien, von dem Jahr 1733 bis 1743* (Göttingen, 1751–52).

4.103n Martigon Lily *Lilium*; see *LP* 1.15.

4.103n alstromeria licta *Alstroemeria ligtu*.

4.103n Memoire's d'Agricult Société royale d'agriculture de Paris, *Mémoires d'agriculture, d'économie rurale et domestique* (Spring 1786): ix.

4.114 Bellis See *LP* 1.69n.

4.114n *prolifera* "*Proliferous*"; having "an umbel [see *LP* 1.373n] more than decompounded [i.e., with leaves subdivided more than twice]" (*SV* 1.xxxix).

4.114n Calendula, marigold See *LP* 2.164.

4.114n Heracium, hawk-weed *Hieracium*; see *LP* 2.163n.

4.121n the materia medica "The remedial substances and preparations used in the practice of medicine; a list of these" (*OED*).

4.121n Mr. Sparman Anders Sparrman (Swedish, 1748–1820), naturalist who accompanied Captain James Cook (English, 1728–79) on his second expedition (1772–75); *A Voyage to the Cape of Good Hope*, 2 vols. (London, 1785).

4.122 harebel See *LP* 1.11.

4.122 primrose See *LP* 1.13.

4.125 Idalian Of the town of Idalium, Cyprus, consecrated to Venus.

4.129 Vulcanian Of Vulcan; see *LP* 3.301–10.

4.141 Masson Mountain in Matlock Dale, Derbyshire; see *LP* 2.87n.

4.144 Derwent See *LP* 2.85.

4.145–58 and n Deep in wide ... its parent flood Darwin rehearsed the arguments he enumerates here in a letter to the Unitarian minister James Pilkington (English, 1752?–1804) dated 5 February 1788 (*Letters* 88–4). While preparing his *View of the Present State of Derbyshire* (Derby, 1789), Pilkington had asked Darwin for information about the thermal springs of the region, whose Roman baths were still popular destinations for travelers. The published work quoted the letter in full. John Whitehurst (see *LP* 3.90n) was among those whom Darwin joined in attributing the springs to "subterraneous fires"; among those who attributed them instead to "the chemical decomposition of pyrites" was Richard Kirwan (Irish, 1733–1812) (qtd. Pilkington 256). See also *EV* 3.166n.

4.145n good thermometers See *LP* 1.280.

4.145n pyrites See *EV* 2.350.

4.145n chalybeate Here, "impregnated or flavored with iron, *esp.* as a mineral water or spring" (*OED*).

4.145n See note in ... of central fires See *AN* 6.

4.155 beds of stalactite "A general term for limestone found in [an icicle-like] formation" (*OED*).

4.158 parent flood Perhaps the River Trent, which the Derwent joins at the Derbyshire-Leicestershire border.

4.161 Fucus See *LP* 1.233.

4.161n Conferva A separate species from *Fucus*; see *LP* 4.270 and *LP* 4.366.

4.161n Dr. Priestly found ... from becoming putrid See *LP* 1.373n.

4.161n Trapa See *LP* 4.170 and n.

4.161n elasticity Power of expansion.

4.162 Echo Garrulous wood-nymph whom Juno doomed never to speak, only to reply. Rejected by the youth Narcissus, she pined away among the mountains until only her voice remained.

4.165–68 So, erst ... the troubled waves An angel "went down at a certain season" into Bethesda, a pool at Jerusalem, "and troubled the water: whosoever then first after the troubling of the water stepped in was made whole of whatsoever disease he had" (John 5.4). Jesus cured a man who had missed his opportunity.

4.170n sisymbrium See *LP* 1.220n.

4.170n petroselinum *Apium petroselinum*, now classed as *Petroselinum crispum*.

4.170n water-parsnip *Sium*.

4.170n rice See *LP* 2.229.

4.170n antient Lotus ... Nymphaea *Nymphaea lotus*; see *LP* 1.177n.

4.170n esculent See *LP* 1.393n.

4.170n Herodotus ... Enterpe Herodotus, *History* (fifth century BCE). Each of the nine books is named after one of the muses; *Euterpe* (music) is Book 2.

4.170n grit stones See *EV* 2.35n.

4.170n white flowered one Presumably *Nymphaea alba*.

4.187 gorget Neck-piece; as armor, a garment or jewelry.

4.189n *Arum* See *LP* 1.131n.

4.189n Tourniff. Joseph Pitton de Tournefort (French, 1656–1708), botanist and physician; *Histoire des plantes qui naissent aux environs de Paris* (Paris, 1698); trans. John Martyn as *Tournefort's History of Plants Growing about Paris* (London, 1732).

4.189n Malpig. Marcello Malpighi (Italian, 1628–94), physician and botanist; *Anatome Plantarum* (London, 1675).

4.189n Dillen. Johann Jakob Dillenius (German, 1687–1747), professor of botany at Oxford; possibly his 3rd ed. of John Ray, *Synopsis methodica stirpium Britannicarum* (London, 1724).

4.189n Rivin. Augustus Quirinus Rivinus (Bachmann) (German, 1652–1723); presumably *Ordo plantarum, quae sunt flore irregulari monopetalo* (Leipzig, 1690).

4.189n spadix "A thick fleshy spike" (see *LP* 1.77n).

4.189n cuticle Epidermis or outermost layer of tissue.

4.193–206 So erst ... as she moves On Dejanira, see *LP* 1.211n; but Darwin is thinking of Queen Omphale, to whom Hercules was enslaved for three years as punishment for killing his friend Iphitus. He wore women's clothing and spun wool, while Omphale wore his Nemean lion's skin. At the conclusion of his service, he spent another three years married to Dejanira.

4.207 Dianthus See *LP* 1.77n.

4.207n There is an ... in No. 32 Daniel Rudberg, *Dissertatio...de peloria* (19 Dec. 1744), *Amoen. acaem.* 1.3 (1749): 55–73; Johan Johansson Haartman (Finnish, 1725–87), *Plantae hybridae* (23 Nov. 1751), *Amoen. acad.* 3.32 (1756): 28–62.

4.207n Urtica alienata Now classed as *Pouzolzia zeylanica.*

4.207n Murray. Syst. Veg. See *LP* 1.131n.

4.207n officinal one ... spiked one *Veronica officinalis* and *Veronica spicata* respectively.

4.207n golden saxifrage *Chrysoplenium.*

4.207n marsh pennywort *Hydrocotyle.*

4.207n Pulteney's view of Linnaeus See *LP* 4.45n; the page cited gives an abstract of Haartman's *Plantae hybridae* (see above).

4.207n Mr. Graberg, Mr. Schreber, and Mr. Ramstrom Johan Martin Gråberg (1741–93), *Fundamentum fructificationes* (16 Oct. 1762), *Amoen. acad.* 6.116 (1761): 279–304; Johann Christian Daniel Schreber (1739–1810), *Theses medicae* (14 June 1760), *Amoen. acad.* 6.103 (1761): 40–43; Christian Ludwig Ramström (1740–82), *Generatio ambigena* (12 Dec. 1759), *Amoen. acad.* 6.101 (1761): 1–16.

4.207n the black cross Typically on the back and shoulders.

4.207n *Amaen. academ.* V. 6. p. 13. Ramström's *Generatio ambigena* (see above).

4.209 heirs Inherits.

4.211–22 So, when the ... the enchanted grove See *LP* 1.288n; but the monstrous birth appears to be Darwin's invention.

4.225 Houries Plural of *houri*, a "nymph of the Muslim Paradise ... applied allusively to a voluptuously beautiful woman" (*OED*).

4.230n The extreme dryness ... from putrid marshes James Lind (Scottish, 1716–94), *An Essay on the Most Effectual Means of Preserving the Health of Seamen in the Royal Navy*, rev. ed. (London, 1774).

4.230n at other seasons ... *Phil. Transact. V. 71* Matthew Dobson (English, 1732–84), "An Account of the Harmattan, a Singular African Wind ... Communicated by John Fothergill" (English, 1712–80), *Phil. Trans.* 71 (1781): 46–57. This paper draws heavily on the observations of Capt. Robert Norris (English, 1724?–91), who made repeated visits to western Africa in the 1770s and would include his own account of the Harmattan in his *Memoirs of the Reign of Bossa Ahádee, King of Dahomy* (London, 1789). The vascular hypothesis is Darwin's.

4.230n fluxes "Morbid or excessive discharge[s]" (*OED*).

4.230n the dry fog ... Hecla, and its neighborhood Darwin is no doubt thinking of the meteorological events of the summer of 1783. The May eruption not of Hecla but of Laki, a volcanic fissure also in Iceland, produced the largest lava flow in recorded history. Sulfurous gases destroyed crops and livestock, resulting in a famine that killed off one-fifth of the Icelandic population. Gilbert White (English, 1720–93) described "the peculiar *haze*, or smokey fog, that prevailed for many weeks in this island, and in every part of *Europe*, and even beyond it's limits ... unlike any thing known within the memory of man"; *The Natural History and Antiquities of Selborne* (London, 1789), 301.

4.230n subsidence Precipitation.

4.233 Atlas Legendary mountain in northwestern Africa; later personified as a king whose domain was called Atlantis, and whom Perseus turned to stone with the head of the Medusa. In many versions of the myth he was condemned to support the heavens on his shoulders, as the mountain was believed to do.

4.234 Gambia Country and river in western Africa.

4.236n Mr. Marsden relates ... Phil. Trans. V. 71. p. 384 William Marsden (Irish, 1754–1836), "Account of a Phenomenon Observed upon the Island of Sumatra ... Communicated by Sir Joseph Banks", *Phil. Trans.* 71 (1781): 383–85. A monsoon is a wind that reverses its direction seasonally; in Sumatra it ordinarily arcs clockwise from the southeast during the "dry" summer months, and clockwise from the northeast during the "wet" winter months.

4.238 zone See *LP* 1.208.

4.247 northern Star See *LP* 2.15.

4.254n The violent cracks ... the ice-mountains Early works of glaciology include Johann Jakob Scheuchzer, *Itinera per Helvaetiae Alpinas regiones* (Leyden, 1723); Gottlieb Sigmund Gruner, *Eisgebirge des Schweitzerlandes*, 3 vols. (Bern, 1760); and Horace-Bénedict de Saussure, *Voyages dans les Alpes*, 4 vols. (Neuchatel, 1779–96).

4.259n See notes on part I See *AN* 32.

4.259n Professor Gmelin ... Account of Siberia See *LP* 4.103n.

4.259n Rail A wading bird.

4.269 Lock-lomond The largest lake in Scotland, along the southern edge of the Highlands.

4.270n Fucus natans See *LP* 3.345n.

4.287–96 So on her ... in her arms Hero, one of Venus' priestesses, loved the youth Leander; she lived in Sestos on the European side of the Hellespont (the present-day Dardanelles), he in Abydos on the Asian side. He swam across the strait to meet her every night, guided by a torch she placed atop a tower. One stormy night, the torch was extinguished and Leander drowned. When Hero discovered his body washed ashore, she flung herself from the tower into the sea.

4.298n stoles Stolons, horizontal branches from the roots of plants.

4.299 asbestus "A mineral of fibrous texture, capable of being woven into an incombustible fabric" (*OED*); see *EV* 2.220.

4.306 mochoes Mochas, decorative stones with branch-like markings; see *EV* 2.215n.

4.306 tessellate "Make into a mosaic" (*OED*).

4.309–26 So, cavern'd round ... the mingled blaze The Wieliczka salt mine, outside Kraków, dates to the thirteenth century. A notable description was given by the geologist Jean-Étienne Guettard (French, 1715–86), "Mémoire sur les mines de sel de Wieliczka en Pologne", *Hist. de l'Acad. roy.* 1762: 493–516.

4.319 Lotta Before destroying Sodom and Gommorah, God permitted Lot to flee to Zoar with his family; but Lot's wife "looked back from behind him, and she became a pillar of salt" (Genesis 19.26).

4.311n these immense masses ... Theory of the Earth See *LP* 3.90n. Hutton says that "subterranean fires" do not merely expose precipitated salt by evaporating seawater, as Darwin mentions, but consolidate it into strata.

4.311n sea-scurvy ... scrophula Now understood as due to a vitamin C deficiency and a bacterial infection, respectively.

4.328n Caprificus Not a Linnaean genus but an older designation, from the Latin, for an inedible variety of *Ficus carica*; see *LP* 3.80.

4.328n Tournefort See *LP* 4.189n; *Relation d'un voyage du Levant*, 3 vols. (Lyon, 1717), trans. *A Voyage to the Levant* (London, 1741).

4.328n Pontedera Giulio Pontedera (Italian, 1688–1757), *Anthologia, sive de floris natura* (Padua, 1720).

4.328n Syst. Nat. See *LP* 1.211n; *Cynips psenes* is the name of the insect.

4.328n Milne's botanical dictionary See *LP* 1.73n; "CAPRIFICATIO ... the very singular husbandry, or management of fig-trees."

4.328n layers *Layer*: "A shoot or twig of a plant fastened down and partly covered with earth, in order that it may strike root while still attached to the parent stock, and so propagate the plant" (*OED*).

4.328n suckers *Sucker*: "A shoot thrown out from the base of a tree or plant, which in most cases may serve for propagation" (*OED*).

4.328n Monsieur de la Hire Jean-Nicolas de la Hire (French, 1685–1727), "Observation sur les figues", *Hist. de l'Acad. roy.* 1712: 275–78.

4.328n barberries See *LP* 1.143n.

4.328n Ischia fig One of several varieties of *Ficus carica*.

4.329 Curculio Weevil.

4.349 Orcades Orkney Islands, northern Scotland.

4.353n cave of Fingal ... island of Staffa Popular attraction in the Inner Hebrides, western Scotland. Fingal, or Finn MacCool (Fionn MacCumhaill), was a legendary Irish warrior; in 1762, James Macpherson (Scottish, 1736–96) published the epic *Fingal*, which he based loosely on genuine Gaelic ballads but claimed to be a translation from the third-century bard Ossian (Oisín), Fingal's son. Although their authenticity was hotly disputed, Macpherson's "Ossianic" productions were an international sensation and figured prominently in the Celtic revival of the eighteenth century, with various nationalist overtones.

4.353n Giants causeway Four-mile formation of hexagonal basalt columns along the northern coast of Antrim, Northern Ireland; legendarily built by Finn MacCool and his giant clansmen as part of a bridge to Staffa. One of the earliest geologists to posit its volcanic rather than oceanic origin—an idea that did not gain immediate acceptance—was Nicolas Desmarest (French, 1725–1815), "Mémoire sur l'origine et la nature du basalte à grandes colonnes polygones", *Hist. de l'Acad. roy.* 1771: 705–75.

4.353n Mr. Keir James Keir (1735–1820), chemist, Lunar Society member and manager of a glassworks at Stourbridge, among other enterprises; "On the Crystallizations Observed in Glass ... Communicated by G. Fordyce", *Phil. Trans.* 66 (1776): 530–42.

4.366n Mr. Ellis "Extract of a Letter from John Ellis [Irish? 1710?–76] ... on the Animal Nature of the Genus of Zoophytes, Called Corallina", 57 (1767): 404–27.

4.369 Dolphin If "scaly", presumably the dorado or mahi-mahi.

4.388n Areoi ... Otaheite Capt. Cook (see *LP* 4.121n) visited Tahiti in the spring of 1769 aboard the *Endeavour*; his journals partly formed the basis for John Hawkesworth's (English, c. 1720–1773) popular but liberty-taking *Account of the Voyages Undertaken...in the Southern Hemisphere*, 3 vols. (London, 1773). Among Hawkesworth's revelations was the existence of a caste called the *areoi* (*arioi*, *arreoy*) that engaged in orgiastic dances (called *timorodee*) and, on occasion, infanticide. The ritual character of these practices was widely misconstrued as sexual license. See also Seward's *Elegy* (see *LP* Interlude 3.50–51), in which Darwin may have had a hand.

Additional Notes

Additional title *Description of the ... N. P. Foersch* See 3.238 and n. The entire note appears as a single article in the *London Magazine*.

Additional 2 Batavia Now Jakarta, northwest Java, Indonesia.

Additional 2 Malayan "Of a people chiefly inhabiting [present-day] Malaysia, Brunei, and parts of Indonesia, and characteristically speaking the Malay language" (*OED*).

Additional 3 *Soura Charta* Surakarta, central Java.

Additional 3 Emperor Pakubuwono III (reigned 1755–88).

Additional 3 *Tinkjoe* Probably Djokjo, i.e. Yogyakarta, south central Java.

Additional 3 Sultan of Java Hamengkubuwono I (reigned 1755–92).

Additional 8 We are told ... for ever uninhabitable Perhaps the Zaqqūm tree, which grows in the depths of Hell and whose poison fruit, shaped like devils' heads, is eaten by the damned; the Koran does not connect it to any specific transgression (37.62–68, 44.43–46).

Additional 12 Massay Perhaps Raden Mas Said, who became Mangkunegara I; see *Additional* 23.

Additional 18 Farriers Those who shoe and treat horses.

Additional 19 *subsultus tendinum* "Twitching of muscles (and their tendons), esp. of the hands and wrist" (*OED*).

Additional 19 *Petechiæ* "A disease characterized by ... small, flat, red or purple spot[s] caused by bleeding into the skin or other organs" (*OED*).

Additional 20 Samarang In central Java.

Additional 21 arrack Alcoholic beverage usually made from fermented rice or coconut.

Additional 23 the last war The crown prince of the kingdom of Mataram (central Java) was Raden Mas Sujana, known as Prince Mangkubumi; he sought the throne from his brother, Pakubuwono II, in the Third Javanese War of Succession in 1746. Mangkubumi was backed by military commander Raden Mas Said; Pakubuwono II, by the VOC. Pakubuwono II fell ill and died, but Mangkubumi continued to be opposed by the VOC and, eventually, by the revolted Said. The conflict was ended under the Treaty of Guyanti in 1755, which divided Mataram into the kingdoms of Surakarta (ruled by Pakubuwono III until 1788) and Yogyakarta (ruled by Mangkubumi, now sultan Hamengkubuwono I, until 1792). Said made a separate peace in 1757 and became Mangkunegara I, ruler of Mangkunegaran within Surakarta until 1795.

Additional 26 Cajoe-Upas Perhaps an alternate name rather than a different species; *upas* is the generic term in Malay for a plant-derived poison.

Additional 26 Macassar In southwestern Sulawesi (Celebes).

Supplement

Supplement 1 Mr. Boothby See *LP* 1.131n; Ashbourne Hall was his family seat.

Supplement 1 Miss Maria Jacson Writer on botany and horticulture (English, 1755–1829); Boothby's cousin.

Supplement 1 R.W. Darwin, Esq. See *LP* Interlude 3.42.

Supplement 3 Newark In east central Nottinghamshire.

Supplement 4 Haggren Lars Christian Haggren, "Om Blommors blickande", *Kongl. Vet.-Acad. Handl.* 9 (1788): 62–64.

Supplement 12 Indian Pink, (*Tagetes patula et erecta*) *SV* associates *Tagetes* with the African or French marigold, Indian pink with *Ipomoea* or *Lonicera*.

Supplement 14 phosphoric Phosphorescent, glowing in the dark.

Supplement 16 *Gen. Evening Post* *General Evening Post* (London), 29 Nov.–2 Dec. 1788: 2.

Table of Plants in *The Loves of the Plants*

Initial Ref. [LP]	*Primary Ref.* [LP]	*Poetic Name* [LP]	*Common Name (species)* [LP, SV]	*Class [Linnaeus]*	*Order [Linnaeus]*	*Genus or Family [Linnaeus]*	*Order [APG III System; * = not an angiosperm]*	*Family (English name) [APG III System]*	*Genus (species) [APG III System]*
Preface para. 35			Shepherd's purse (bursa-pastoris)	Thlaspi	Tetradynamia	Siliculosa	Brassicales	Brassicaceae (cabbage)	Capsella (bursa-pastoris)
Preface para. 35			Wall-flower	Tetradynamia	Siliquosa	Cheiranthus	Brassicales	Brassicaceae (cabbage)	Erysimum
1.11	4.103	Galantha	Bulbous violet, snowdrop (nivalis)	Hexandria	Monogynia	Galanthus	Asparagales	Amaryllidaceae (amaryllis)	Galanthus (nivalis)
1.11			Hairbell, hyacinth, jacinth	Hexandria	Monogynia	Hyacinthus	Asparagales	Asparagaceae	Hyacinthus
1.13	1.213		Heart's ease, pansy, three faces under a hood, trinity-herb, violet	Syngenesia	Monogamia	Viola	Malpighiales	Violaceae	Viola
1.13			Auricula, cowslip, bear's-ear, ox-lips, pagils or paigles, primrose	Pentandria	Monogynia	Primula	Ericales	Primulaceae (primrose)	Primula
1.15	4.103n		Lily, martagon, Turk's cap	Hexandria	Monogynia	Lilium	Liliales	Liliaceae (lily)	Lilium
1.17			Briar, hep-tree, rose	Icosandria	Polygynia	Rosa	Rosales	Rosaceae (rose)	Rosa
1.19	1.211	Lonicera	Dwarf cherry, fly honeysuckle, honeysuckle (caprifolium), Indian pink, snowberry-bush, woodbind or woodbine	Pentandria	Monogynia	Lonicera	Dipsacales	Caprifoliaceae (honeysuckle)	Lonicera (caprifolium)

(Continued)

Initial Ref. [LP]	*Primary Ref.* [LP]	*Poetic Name* [LP]	*Common Name (species)* [LP, SV]	*Class [Linnaeus]*	*Order [Linnaeus]*	*Genus or Family [Linnaeus]*	*Order [APG III System; * = not an angiosperm]*	*Family (English name) [APG III System]*	*Genus (species) [APG III System]*
1.39		Canna	Cane, indian reed (indica), indian shot, plantain shot	Monandria	Monogynia	Canna	Rosales	Cannaceae	Canna (indica)
1.45		Callitriche	Fine-hair, stargrass, starry duck's-meat	Monandria	Digynia	Callitriche	Lamiales	Plantaginaceae (plantain)	Callitriche
1.51		Collin/ Collinia		Diandria	Monogynia	Collinsonia	Lamiales	Lamiaceae (mint, deadnettle)	Collinsonia
1.51n			Devil in the bush, flower fennel	Polyandria	Pentagynia	Nigella	Ranunculales	Ranunculaceae (buttercup, crowfoot)	Nigella
1.51n			Codlins in cream, French willow, herb willow, podded loose-strife, rose bay willow herb (angustifolium), willow herb	Octandria	Monogynia	Epilobium	Myrtales	Onagraceae (willowherb, evening primrose)	Chamerion (angustifoli-um)
1.57		Genista	Dyer's broom, dyer's weed, dwarf broom, single-seeded broom	Diadelphia	Decandria	Genista	Fabales	Fabaceae (legume)	Genista
1.57			All-spice, Jamaica pepper, myrtle, pimento	Icosandria	Monogynia	Myrtus	Myrtales	Myrtaceae (myrtle)	Myrtus
1.57n			Beesom (scoparium), broom	Diadelphia	Decandria	Spartium	Fabales	Fabaceae (legume)	Cytisus (scoparius)
1.60		Melissa	Balm or baulm, calamint	Didynamia	Gymnospermia	Melissa	Lamiales	Lamiaceae (mint, deadnettle)	Melissa

1.60n			Black horehound	Didynamia	Gymnospermia	Ballota	Lamiales	Lamiaceae (mint, deadnettle)	Ballota
1.61		Meadia	American cowslip (meadia)	Pentandria	Monogynia	Dodecatheon	Ericales	Primulaceae (primrose)	Dodecatheon (meadia)
1.61n			Borage	Pentandria	Monogynia	Borago	not placed	Boraginaceae (borage, forget-me-not)	Borago
1.61n	3.311	Cyclamen	Sow-bread	Pentandria	Monogynia	Cyclamen	Ericales	Primulaceae (primrose)	Cyclamen
1.61n			Batchelor's pear, eggplant, felon-wort, love apple, mad apple, potato, winter cherry, wolf's peach	Pentandria	Monogynia	Solanum	Solanales	Solanaceae (nightshade)	Solanum
1.65		Curcuma	Turmeric or turmerick	Monandria	Monogynia	Curcuma	Zingiberales	Zingiberaceae (ginger)	Curcuma
1.65n	2.67	Lina	All-seed, flax, least rupture-wort	Pentandria	Pentagynia	Linum	Malpighiales	Linaceae	Linum
1.65n			Simpler's joy, vervain	Diandria	Monogynia	Verbena	Lamiales	Vervenaceae (verbena, vervain)	Verbena
1.65n				Hexandria	Monogynia	Albuca	Asparagales	Asparagaceae	Albuca
1.65n			Milk wood, trumpet-flower, tulip-flower, white wood, yellow jasmine	Didynamia	Angiospermia	Bignonia	Lamiales	Bignoniaceae (bignonias)	Catalpa (bignonioides)
1.65n			Hedge hyssop	Diandria	Monogynia	Gratiola	Lamiales	Plantaginaceae (plantain)	Gratiola
1.65n			Hemlock-leaved geranium, water hemlock (cicutarium)	Monadelphia	Decandria	Geranium	Geraniales	Geraniaceae	Erodium (cicutarium)

(*Continued*)

(Continued)

Initial Ref. [LP]	*Primary Ref.* [LP]	*Poetic Name* [LP]	*Common Name (species)* [LP, SV]	*Class [Linnaeus]*	*Order [Linnaeus]*	*Genus or Family [Linnaeus]*	*Order [APG III System; * = not an angiosperm]*	*Family (English name) [APG III System]*	*Genus (species) [APG III System]*
1.65n	1.191	Helianthus	Jerusalem artichoke, sunflower	Syngenesia	Polygamia frustranea	Helianthus	Asterales	Asteraceae (aster, sunflower)	Helianthus
1.65n	1.69n		Gelder rose, gelderland rose, marsh elder, pliant mealy-tree, snowball-tree	Pentandria	Trigynia	Viburnum	Dipsacales	Adoxaceae	Viburnum
1.69		Alcea	Hollyhock, rose mallow	Monadelphia	Polyandria	Alcea	Malvales	Malvaceae (mallow)	Alcea
1.69n			Bell-flower, Canterbury bells, common esculent rampions, Coventry bells, throat-wort	Pentandria	Monogynia	Campanula	Asterales	Campanulaceae (bellflower)	Campanula
1.69n			Thorn apple (stramonium)	Pentandria	Monogynia	Datura	Solanales	Solanaceae (nightshade)	Datura (stramonium)
1.69n			Marsh marigold	Polyandria	Monogynia	Caltha	Ranunculales	Ranunculaceae (buttercup, crowfoot)	Caltha
1.69n			Paeony or peony	Polyandria	Digynia	Paeonia	Saxifragales	Paeoniaceae	Paeonia
1.69n			Lark's heel, lark's spur, lauskraut, royal consound	Polyandria	Trigynia	Delphinium	Ranunculales	Ranunculaceae (buttercup, crowfoot)	Delphinium
1.69n			Columbine	Polyandria	Pentagynia	Aquilegia	Ranunculales	Ranunculaceae (buttercup, crowfoot)	Aquilegia
1.69n	4.114	Bellis	Daisy, gowan, lesser consound	Syngenesia	Polygamia superflua	Bellis	Asterales	Asteraceae (aster, sunflower)	Bellis
1.69n			Feverfew	Syngenesia	Polygamia superflua	Matricaria	Asterales	Asteraceae (aster, sunflower)	Matricaria

1.69n	1.284	Anemone	Hepatica, noble liver-wort, pasque-flower, wind-flower, wood anemone	Polygamia	Polygynia	Anemone	Ranunculales	Ranunculaceae (buttercup, crowfoot)	Anemone
1.69n			(Water) crow-foot (aquatilis), lesser celandine, pile-wort, spear-wort, Turk's turban	Polyandria	Polygynia	Ranunculus	Ranunculales	Ranunculaceae (buttercup, crowfoot)	Ranunculus
1.69n	2.268	Papaver	Corn rose, poppy	Polyandria	Monogynia	Papaver	Ranunculales	Papaveraceae (poppy)	Papaver
1.71		Iris	Flag, flag-flower, flower de luce, Spanish nut, stinking gladwin	Triandria	Monogynia	Iris	Asparagales	Iridaceae	Iris
1.73		Cupressus	Cypress	Monoecia	Monadelphia	Cupressus	Pinales*	Cupressaceae (cypress)	Cupressus
1.73n	3.191	Urtica	Nettle, Spanish marjoram	Monoecia	Tetrandria	Urtica	Rosales	Urticaceae (nettle)	Urtica
1.73n			Cedar of Libanus, fir, frankincense tree, larch-tree (larix), pinaster, pine-tree, pitch-tree	Monoecia	Monadelphia	Pinus	Pinales*	Pinaceae (pine)	Larix
1.73n			Cedar (bermudiana), Juniper, savin (sabina)	Dioecia	Monadelphia	Juniperus	Pinales*	Cupressaceae (cypress)	Juniperus
1.75		Osyris	Poet's rosemary	Dioecia	Triandria	Osyris	Santalales	Santalaceae (sandalwood)	Osyris

(Continued)

(Continued)

Initial Ref. [LP]	*Primary Ref.* [LP]	*Poetic Name* [LP]	*Common Name (species)* [LP, SV]	*Class [Linnaeus]*	*Order [Linnaeus]*	*Genus or Family [Linnaeus]*	*Order [APG III System; * = not an angiosperm]*	*Family (English name) [APG III System]*	*Genus (species) [APG III System]*
1.77		Plantago	Bucks-horn plantain, flea-wort, hart's-horn plantain, plantain, ribwort, rose-plantain (rosea)	Tetrandria	Monogynia	Plantago	Lamiales	Plantaginaciae (plantain)	Plantago
1.77n				Syngenesia	Polygamia superflua	Xeranthemum	Asterales	Asteraceae (aster, sunflower)	Xeranthemum
1.77n	4.207	Dianthus	Carnation, clove July-flower, Fairchild's mule (superbus x Caryophyllus aromaticus), old man's head, painted ladies, pink, piquets, proliferous pink (prolifer), proud pink (superbus), sweet John, sweet William	Decandria	Digynia	Dianthus	Caryophyllales	Caryophyllaceae (pink or carnation)	Dianthus (superbus)
1.77n			Osier, sallow, willow	Dioecia	Diandria	Salix	Malpighiales	Salicaceae (willow)	Salix
1.86		Anthoxa	Vernal grass	Diandria	Digynia	Anthoxanthum	Poales	Poaceae (true grasses)	Anthoxanthum
1.86n			(Bush) fescue (dumetorum)	Triandria	Digynia	Festuca	Poales	Poaceae (true grasses)	Festuca (rubra)
1.86n			Ciboules, cives, eschalot, garlick, (magic) onion (magicum), ramsons, rocambole, scallion	Hexandria	Monogynia	Allium	Asparagales	Amaryllidaceae (amaryllis)	Allium (nigrum)

1.86n		Adder's-wort, (viviparous) bistort (viviparum), buck wheat, knot grass, lakeweed, mangrove grape, persicaria	Octandria	Trigynia	Polygonum	Caryophyllales	Polygonaceae (knotweed, smartweed, buckwheat)	Persicaria (vivipara)
1.86n		Birch	Monoecia	Tetrandria	Betula	Fagales	Betulaceae (birch)	Betula
1.88		Furze, gorse, scorpion's thorn	Diadelphia	Decandria	Ulex	Fabales	Fabaceae (legume)	Ulex
1.93	Osmunda	Osmund royal	Cryptogamia	Filices	Osmunda	Osmundales*	Osmundaceae (royal fern)	Osmunda
1.97	Chondrilla	Gum succory	Syngenesia	Polygamia aequalis	Chondrilla	Asterales	Asteraceae (aster, sunflower)	Chondrilla
1.97n		Blue-bonnet, blue-bottle, bottle-flower, centaury, knapweed, star-thistle, sultan-flower, sweet sultan, thorny knapweed	Syngenesia	Polygamia frustranea	Centaurea	Asterales	Asteraceae (aster, sunflower)	Centaurea
1.97n		Artichoke, chardon	Syngenesia	Polygamia aequalis	Cynara	Asterales	Asteraceae (aster, sunflower)	Cynara
1.97n		Globe-thistle	Syngenesia	Polygamia segregata	Echinops	Asterales	Asteraceae (aster, sunflower)	Echinops

(*Continued*)

(Continued)

Initial Ref. [LP]	*Primary Ref.* [LP]	*Poetic Name* [LP]	*Common Name (species)* [LP, SV]	*Class [Linnaeus]*	*Order [Linnaeus]*	*Genus or Family [Linnaeus]*	*Order [APG III System; * = not an angiosperm]*	*Family (English name) [APG III System]*	*Genus (species) [APG III System]*
1.108		Lychnis	Batchelor's buttons, flower of Bristol, flower of Constantinople, campion, Jerusalem cross, knights cross, scarlet cross, nonsuch, ragged robin	Decandria	Pentagynia	Lychnis	Caryophyllales	Caryophyllaceae (pink or carnation)	Lychnis
1.119		Gloriosa	Superb lily (superba)	Hexandria	Monogynia	Gloriosa	Liliales	Colchicaceae	Gloriosa
1.119n			Chequer'd tulip, fritillary, monster	Hexandria	Monogynia	Fritillaria	Liliales	Liliaceae (lily)	Fritillaria (persica)
1.119n			Poley grass, purple loose-strife (salicaria), spiked loose-strife, willlow herb	Dodecandria	Monogynia	Lythrum	Myrtales	Lythraceae	Lythrum (salicaria)
1.119n			Hollow root	Octandria	Tetragynia	Adoxa	Dipsacales	Adoxaceae	Adoxa
1.119n			Break-stone, cushion ladies, none so pretty, rue-leaved whitlow grass, saxifrage, sedum pyramidal	Decandria	Digynia	Saxifraga	Saxifragales	Saxifragaceae	Saxifraga
1.131		Silene	Catchfly, fly bane, fly-wort, viscous campion	Decandria	Trigynia	Silene	Caryophyllales	Caryophyllaceae (pink or carnation)	Silene
1.131n			Campion, spatling poppy, white behen	Decandria	Trigynia	Cucubalus	Caryophyllales	Caryophyllaceae (pink or carnation)	Silene (otites)
1.131n			Fly-trap (muscipula)	Decandria	Monogynia	Dionaea	Caryophyllales	Droseraceae (sundew)	Dionaea (muscipula)
1.131n	4.189	Ara	Cuckow-pint, dragons, fryer's cowl, lords and ladies, wake robin	Gynandria	Polyandria	Arum	Alismatales	Araceae (arum)	Arum

1.131n	1.333	Dypsaca	Teasel	Tetrandria	Monogynia	Dipsacus	Dipsacales	Caprifoliaceae (honeysuckle); includes Dipsacaceae (teasel)	Dipsacus
1.131n	1.199	Drosera	Sun-dew	Pentandria	Pentagynia	Drosera	Caryophyllales	Droseraceae (sundew)	Drosera
1.143		Ilex	False tea, holly, hop-tree, laurel-leaved bladder nut, Paraguay tea, Saint Bartholemew's herb, South-sea tea, yapon	Tetrandria	Tetragynia	Ilex	Aquifoliales	Aquifoliaceae (holly)	Ilex
1.143n			Barberry or berberry, piperidge bush	Hexandria	Monogynia	Berberis	Ranunculales	Berberidaceae (barberry)	Berberis
1.143n			Haw, hawthorn, May bush	Icosandria	Digynia	Crataegus	Rosales	Rosaceae (rose)	Crataegus
1.143n			Curran tree, gooseberry	Pentandria	Monogynia	Ribes	Saxifragales	Grossulariaceae	Ribes
1.157		Kleinhovia		Gynandria	Decandria	Kleinhovia	Malvales	Malvaceae (mallow)	Kleinhovia
1.157n			African calabash tree, Ethiopian sour-gourd	Monadelphia	Polyandria	Adansonia	Malvales	Malvaceae (mallow)	Adansonia
1.177		Tulipa	Tulip	Hexandria	Monogynia	Tulipa	Liliales	Liliaceae (lily)	Tulipa
1.177n	2.163	Nymphaea	Water-lily	Polyandria	Monogynia	Nymphaea	Nymphaeales	Nymphaeaceae (water lilies)	Nymphaea (alba)
1.177n			Louse-wort, rattle	Didynamia	Angiospermia	Pedicularis	Lamiales	Orobanchaceae	Pedicularis (hirsuta)

(Continued)

(Continued)

Initial Ref. [LP]	*Primary Ref.* [LP]	*Poetic Name* [LP]	*Common Name (species)* [LP, SV]	*Class [Linnaeus]*	*Order [Linnaeus]*	*Genus or Family [Linnaeus]*	*Order [APG III System; * = not an angiosperm]*	*Family (English name) [APG III System]*	*Genus (species) [APG III System]*
1.177n			Flax-leaved laurel, mezereon (mezereum), spurge laurel (laureola), spurge olive, tarton-raire (tartonraira)	Octandria	Monogynia	Daphne	Malvales	Thymelaeaceae	Daphne (mezereum)
1.184		Colchica	(Autumnal) meadow-saffron (autumnale)	Hexandria	Trigynia	Colchicum	Liliales	Colchicaceae	Colchicum
1.184n			Ash	Polygamia	Dioecia	Fraxinus	Lamiales	Oleaceae	Fraxinus
1.184n			Black Virginian pistacia-tree, hazel-leaved pistacia, witch hazel	Tetrandria	Monogynia	Hamamelis	Saxifragales	Hamamelidaceae (witch-hazel)	Hamamelis
1.194n	1.226	Visca	Mistletoe	Dioecia	Tetrandria	Viscum	Santalales	Santalaceae (sandalwood)	Viscum
1.194n			Cleavers, clevers, clivers, goose-grass, goose-grease (aparine); bedstraw, cheese rennet, lady's bedstraw	Tetrandria	Monogynia	Galium	Gentianales	Rubiaciae (coffee, madder, bedstraw)	Galium (aparine)
1.194n			Burdock	Syngenesia	Polygamia aequalis	Arctium	Asterales	Asteraceae (aster, sunflower)	Arctium
1.194n			Houndstongue	Pentandria	Monogynia	Cynoglossum	not placed	Boraginaceae (borage, forget-me-not)	Cynoglossum
1.194n	4.73	Avena	Oats	Triandria	Digynia	Avena	Poales	Poaceae (true grasses)	Avena
1.194n	3.131	Impatiens	Balsam, female balsamine, immortal eagle flower, noli me tangere, touch me not	Syngenesia	Monogamia	Impatiens	Ericales	Balsaminaceae (balsam)	Impatiens

1.200			Rush	Hexandria	Monogynia	Juncus	Poales	Juncaceae (rush)	Juncus
1.211n			Aconite, monk's-hood, wolf's bane	Polyandria	Trigynia	Aconitum	Ranunculales	Ranunculaceae (buttercup, crowfoot)	Aconitum
1.211n	2.199	Helleboras	Hellebore, Christmas rose (niger), winter aconite, winter wolf's bane	Polyandria	Polygynia	Helleborus	Ranunculales	Ranunculaceae (buttercup, crowfoot)	Helleborus
1.211n			Bee-orchis, bird's nest, fly-orchis, triple lady's traces, tway blade, twy blade	Gynandria	Diandria	Ophrys	Asparagales	Orchidaceae (orchid)	Ophrys
1.220		Draba	Whitlow-grass	Tetradynamia	Siliculosae	Draba	Brassicales	Brassicaceae (cabbage)	Draba
1.220n			Mustard	Tetradynamia	Siliquosa	Sinapis	Brassicales	Brassicaceae (cabbage)	Sinapis
1.220n			Marsh rocket, water-cress (nasturtium), water radish, water rocket, winter rocket	Tetradynamia	Siliquosa	Sisymbrium	Brassicales	Brassicaceae (cabbage)	Rorippa (nasturtium-aquaticum)
1.220n			Bore-cole, broccoli, cabbage, cale, cauliflower, cole-rape, cole-seed, cole-wort, French turnep, rape, savoys, turnep (rapa)	Tetradynamia	Siliquosa	Brassica	Brassicales	Brassicaceae (cabbage)	Brassica

(*Continued*)

(Continued)

Initial Ref. [LP]	*Primary Ref.* [LP]	*Poetic Name* [LP]	*Common Name (species)* [LP, SV]	*Class [Linnaeus]*	*Order [Linnaeus]*	*Genus or Family [Linnaeus]*	*Order [APG III System; * = not an angiosperm]*	*Family (English name) [APG III System]*	*Genus (species) [APG III System]*
1.226n			Wild pine	Hexandria	Monogynia	Tillandsia	Poales	Bromeliaceae (bromeliads)	Tillandsia
1.232		Zostera	Grass-wrack	Gynandria	Polyandria	Zostera	Alismatales	Zosteraceae (seagrass)	Zostera
1.232n			Hyacinth lily, Peruvian hyacinth, sea onion, squill, starry hyacinth	Hexandria	Monogynia	Scilla	Asparagales	Asparagaceae	Scilla
1.233	4.161	Fucus	Sea weed, wrack	Cryptogamia	Algae	Fucus	Fucales	Fucaceae	Fucus
1.250		Barometz	Common fern, rough spleen-wort, Tartarian lamb (barometz)	Cryptogamia	Filices	Polypodium	Cyatheales*	Dickinsoniaceae	Cibotium (barometz)
1.253			Mastich thyme, mother of thyme, stone basil, thyme, wild basil	Didynamia	Gymnospermia	Thymus	Lamiales	Lamiaceae (mint, deadnettle)	Thymus
1.267		Mimosa	Acacia, binding bean tree, Egyptian thorn, humble plant, sensitive plant, spunge-tree	Polygamia	Monoecia	Mimosa	Fabales	Fabaceae (legume)	Mimosa
1.295		Lichen	Liver-wort (calcareum), moss-tree, tree moss	Cryptogamia	Algae	Lichen	various*	various	various
1.313n			Banana, plantain-tree	Polygamia	Monoecia	Musa	Zingiberales	Musaceae	Musa
1.313n				Gynandria	Tetrandria	Nepenthes	Caryophyllales	Nepenthaceae	Nepenthes
1.341		Rubia	Madder	Tetrandria	Monogynia	Rubia	Gentianales	Rubiaciae (coffee, madder, bedstraw)	Rubia
1.361		Vallisner		Dioecia	Diandria	Vallisneria	Alismatales	Hydrocharitaceae (tape-grass)	Vallisneria (spiralis)
1.373		Ulva	Laver, sloke	Cryptogamia	Algae	Ulva	Ulvales*	Ulvaceae	Ulva
1.373n			Bastard senna, bladder-senna	Diadelphia	Decandria	Colutea	Fabales	Fabaceae (legume)	Colutea

1.373n			Bladder-nut	Pentandria	Trigynia	Staphylea	Crossosomatales	Staphyleaceae	Staphylea
1.373n			Ketmia	Monadelphia	Polyandria	Hibiscus	Malvales	Malvaceae (mallow)	Hibiscus (trionum)
1.393		Tremella	Star-jelly (nostoc)	Cryptogamia	Algae	Tremella	Tremellales*	Tremellaceae	Tremella
1.393n			Wheat	Triandria	Digynia	Triticum	Poales	Poaceae (true grasses)	Triticum
1.393n			(Fly) agaric (muscarum), mushrooms, paddock-stool	Cryptogamia	Fungi	Agaricus	Agaricales*	Agaricaceae	Agaricus
1.393n			(Immodest) stink-horn (impudicus)	Cryptogamia	Fungi	Phallus	Phallales	Phallaceae	Phallus (impudicus)
1.437	3.40	Laura	Apricot (armeniaca), azerira, bird-cherry, black thorn, bullace tree, cherry (cerasus), cherry laurel, damson tree, German acacia, laurel (laurocerasus), mahaleb, plum-tree, prune, sloe-tree	Icosandria	Monogynia	Prunus	Rosales	Rosaceae (rose)	Prunus
2.7		Carlina	Carline thistle	Syngenesia	Polygamia aequalis	Carlina	Asterales	Asteraceae (aster, sunflower)	Carlina
2.87		Gossypia	Cotton	Monadelphia	Polyandria	Gossypium	Malvales	Malvaceae (mallow)	Gossypium
2.87n			Hemp	Dioecia	Pentandria	Cannabis	Rosales	Cannabaceae	Cannabis
2.87n			Fustic tree, Mulberry-tree	Monoecia	Triandria	Morus	Rosales	Moraceae (mulberry, fig)	Morus
2.105		Papyra	Galingale	Triandria	Monogynia	Cyperus	Poales	Cyperaceae (sedge)	Cyperus (papyrus)
2.163		Lapsana	Nipple-wort	Syngenesia	Polygamia aequalis	Lapsana	Asterales	Asteraceae (aster, sunflower)	Lapsana

(Continued)

(Continued)

Initial Ref. [LP]	*Primary Ref.* [LP]	*Poetic Name* [LP]	*Common Name (species)* [LP, SV]	*Class [Linnaeus]*	*Order [Linnaeus]*	*Genus or Family [Linnaeus]*	*Order [APG III System; * = not an angiosperm]*	*Family (English name) [APG III System]*	*Genus (species) [APG III System]*
2.163n			Dandelion, monk's-head, yellow devil's-bit	Syngenesia	Polygamia aequalis	Leontodon	Asterales	Asteraceae (aster, sunflower)	Taraxacum (officinale)
2.163n			Creeping mouse-ear, golden lung-wort, golden mouse-ear, (mouse-eared) hawkweed (pilosella)	Syngenesia	Polygamia aequalis	Hieracium	Asterales	Asteraceae (aster, sunflower)	Hieracium (pilosella)
2.163n			Hare's lettuce, sow-thistle	Syngenesia	Polygamia aequalis	Sonchus	Asterales	Asteraceae (aster, sunflower)	Sonchus
2.163n			(Cultivated) lettice (sativa)	Syngenesia	Polygamia aequalis	Lactuca	Asterales	Asteraceae (aster, sunflower)	Lactuca
2.163n			Go to bed at noon, goat's-beard	Syngenesia	Polygamia aequalis	Tragopogon	Asterales	Asteraceae (aster, sunflower)	Tragopogon
2.163n			(Tawny) day-lily (fulva), lily asphodel, St. Bruno's lily	Hexandria	Monogynia	Hemerocallis	Asparagales	Xanthorrhoeaceae	Hemerocallis
2.163n			Bindweed, Indian turbith, sea cole-wort, Spanish potatoe, Syrian scammony, turbith of the shops	Pentandria	Monogynia	Convolvulus	Solanales	Convolvulaceae (bindweed, morning glory)	Convolvulus
2.163n			Mallow, vervain mallow	Monadelphia	Polyandria	Malva	Malvales	Malvaceae (mallow)	Malva
2.163n			Sandwort	Decandria	Trigynia	Arenaria	Caryophyllales	Caryophyllaceae (pink or carnation)	Arenaria
2.163n			Pimpernel	Pentandria	Monogynia	Anagallis	Ericales	Primulaceae (primrose)	Anagallis

2.163n			Purslane	Diodecandria	Monogynia	Portulaca	Caryophyllales	Portulacaceae (purslane)	Portulaca
2.163n			Succory	Syngenesia	Polygamia aequalis	Cichoreum	Asterales	Asteraceae (aster, sunflower)	Cichorium
2.163n				Syngenesia	Polygamia aequalis	Hypochaeris	Asterales	Asteraceae (aster, sunflower)	Hypochaeris
2.163n			Bastard hawk-weed	Syngenesia	Polygamia aequalis	Crepis	Asterales	Asteraceae (aster, sunflower)	Crepis
2.163n			Ox-tongue	Syngenesia	Polygamia aequalis	Picris	Asterales	Asteraceae (aster, sunflower)	Picris
2.164		Calendula	(Field) marigold (arvensis)	Syngenesia	Polygamia necessaria	Calendula	Asterales	Asteraceae (aster, sunflower)	Calendula
2.227		Menispermum	Indian berry (cocculus)	Dioecia	Dodecandria	Menispermum	Ranunculales	Menispermaceae	Menispermum
2.229			Rice	Hexandria	Digynia	Oryza	Poales	Poaceae (true grasses)	Oryza
2.266			(Branchy) asphodel (ramosus)	Hexandria	Monogynia	Asphodelus	Asparagales	Xanthorrhoeaceae	Asphodelus (ramosus)
2.266			Amaranth, blite, flower gentle, love lies a-bleeding, prince's feather	Monoecia	Pentandria	Amaranthus	Caryophyllales	Amaranthaceae (amaranth)	Amaranthus
2.301		Cista	Rock-rose	Polyandria	Monogynia	Cistus	Malvales	Cistaceae (rock rose)	Cistus (ladanifer)
2.301n			Night primrose, primrose-tree	Octandria	Monogynia	Oenothera	Myrtales	Onagraceae (willowherb, evening primrose)	Oenothera

(*Continued*)

(Continued)

Initial Ref. [LP]	*Primary Ref.* [LP]	*Poetic Name* [LP]	*Common Name (species)* [LP, SV]	*Class [Linnaeus]*	*Order [Linnaeus]*	*Genus or Family [Linnaeus]*	*Order [APG III System; * = not an angiosperm]*	*Family (English name) [APG III System]*	*Genus (species) [APG III System]*
2.345		Cinchona	True Jesuit's bark-tree	Pentandria	Monogynia	Cinchona	Gentianales	Rubiaceae (coffee, madder, bedstraw)	Cinchona
2.421		Digitalis	Foxglove	Didynamia	Angiospermia	Digitalis	Lamiales	Plantaginaceae (plantain)	Digitalis
3.7		Circaea	Enchanter's nightshade	Diandria	Monogynia	Circaea	Myrtales	Onagraceae (willowherb, evening primrose)	Circaea
3.7n			(White) bryony (alba), white vine	Monoecia	Syngenesia	Bryonia	Cucurbitales	Cucurbitaceae (gourd)	Bryonia (alba)
3.7n			Mandrake	Pentandria	Monogynia	Mandragora	Solanales	Solanaceae (nightshade)	Mandragora
3.7n			Cob-nut, filberd, hazel nut, hazle	Monoecia	Polyandria	Corylus	Fagales	Betulaceae (birch)	Corylus
3.19			Yew	Dioecia	Polyandria	Taxus	Pinales*	Taxaceae (yew)	Taxus
3.40n			Almond, peach (persica)	Icosandria	Monogynia	Amygdalus	Rosales	Rosaceae (rose)	Prunus (persica)
3.40n			Walnut	Monoecia	Polyandria	Juglans	Fagales	Juglandaceae (walnut)	Juglans
3.74n			Tuberose (tuberosa)	Hexandria	Monogynia	Polianthes	Asparagales	Asparagaceae	Polianthes (tuberosa)
3.81		Fica	Banian tree, fig, Indian fig-tree (indica), Indian God tree, Pharaoh's fig, sycamore	Polygamia	Trioecia	Ficus	Rosales	Moraceae (mulberry, fig)	Ficus
3.131n			Crane's bill, dove's foot, herb Robert	Monadelphia	Pentandria	Geranium	Geraniales	Geraniaceae	Geranium
3.131n			Barley	Triandria	Digynia	Hordeum	Poales	Poaceae (true grasses)	Hordeum
3.184		Dictamna	Fraxinella, white dittany	Decandria	Monogynia	Dictamnus	Sapindales	Rutaceae (rue, citrus)	Dictamnus

3.184n		Oroonoka, tobacco, yellow henbane	Pentandria	Monogynia	Nicotiana	Solanales	Solanaceae (nightshade)	Nicotiana
3.184n		Ague tree, alligator pear, avocado pear, bay (nobilis), Benjamin tree, camphire tree (camphora), cinnamon (cinnamomum), sassafras-tree, spice wood, white cinnamon, winter's bark	Enneandria	Monogynia	Laurus	Laurales	Lauraceae (laurel)	Cinnamomum
3.184n		Balsam of Peru	Decandria	Monogynia	Myroxylon	Fabales	Fabaceae (legume)	Myroxylon
3.188	Mancinella	Manchineel-tree (mancinella)	Monoecia	Monadelphia	Hippomane	Malpighiales	Euphorbiaciae (spurge)	Hippomane
3.188n		Deadly nightshade (belladonna), dwale	Pentandria	Monogynia	Atropa	Solanales	Solanaceae (nightshade)	Atropa (belladonna)
3.188n		Henbane	Pentandria	Monogynia	Hyoscyamus	Solanales	Solanaceae (nightshade)	Hyoscyamus
3.193	Lobelia	Cardinal-flower, crested rampions, water gladiole	Syngenesia	Monogamia	Lobelia	Asterales	Campanulaceae (bellflower)	Hippobroma (longiflora)
3.225n		Nutmeg	Monoecia	Monandria	Myristica	Magnoliales	Myristicaceae (nutmeg)	Myristica
3.259	Cuscuta	Dodder, dodder of thyme	Tetrandria	Digynia	Cuscuta	Solanales	Convolvulaceae (bindweed, morning glory)	Cuscuta

(Continued)

(Continued)

Initial Ref. [LP]	*Primary Ref.* [LP]	*Poetic Name* [LP]	*Common Name (species)* [LP, SV]	*Class [Linnaeus]*	*Order [Linnaeus]*	*Genus or Family [Linnaeus]*	*Order [APG III System; * = not an angiosperm]*	*Family (English name) [APG III System]*	*Genus (species) [APG III System]*
3.259n			Climbing five-leaved vine of Canada, ivy, Virginian creeper	Pentandria	Monogynia	Hedera	Apiales	Araliaceae (Aralia or ivy)	Hedera
3.259n			Lady's bower, old man's beard, traveller's joy, virgin's bower	Polyandria	Polygynia	Clematis	Ranunculales	Ranunculaceae (buttercup, crowfoot)	Clematis
3.259n			Hop	Dioecia	Pentandria	Humulus	Rosales	Cannabaceae	Humulus
3.259n			Black bindweed, black bryony, black vine, lady's seal	Dioecia	Hexandria	Tamus	Dioscoreales	Dioscoreaceae	Dioscorea
3.259n			French bean, kidney bean	Diadelphia	Decandria	Phaseolus	Fabales	Fabaceae (legume)	Phaseolus
3.259n			Malabar nightshade	Pentandria	Trigynia	Basella	Caryophyllales	Basellaceae	Basella
3.259n				Pentandria	Digynia	Cynanchum	Gentianales	Apocynaceae (dogbane)	Cynanchum
3.259n			Bastard spurge, burning thorny plant, Medusa's head, milk-wort, poison bush, spurge, wart-wort	Dodecandria	Trigynia	Euphorbia	Malpighiales	Euphorbiaciae (spurge)	Euphorbia
3.259n			Hemp-agrimony	Syngenesia	Polygamia aequalis	Eupatorium	Asterales	Asteraceae (aster, sunflower)	Eupatorium
3.259n			Millett, panic, panic-grass	Triandria	Digynia	Panicum	Poales	Poaceae (true grasses)	Panicum (brevifolium)
3.259n				Hexandria	Monogynia	Asparagus	Asparagales	Asparagaceae	Asparagus
3.287		Vitis	Vine	Pentandria	Monogynia	Vitis	Vitales	Vitaceae	Vitis
3.311n			Clover, Dutch clover, honeysuckle grass, trefoil	Diadelphia	Decandria	Trifolium	Fabales	Fabaceae (legume)	Trifolium

3.345	Cassia	Bastard senna, French guava, pudding pipe-tree, senna of the shops, wild senna	Decandria	Monogynia	Cassia	Fabales	Fabaceae (legume)	Cassia
3.345n		Acajou, cashew-nut	Enneandria	Monogynia	Anacardium	Sapindales	Anacardiaceae (cashew, sumac)	Anacardium
3.345n		Citrul, (flaggon) gourd (lagenaria), pompion, squash, water melon	Monoecia	Syngenesia	Cucurbita	Cucurbitales	Cucurbitaceae (gourd)	Lagenaria (siceraria)
3.345n			Diadelphia	Decandria	Piscidia	Fabales	Fabaceae (legume)	Piscidia (piscipula)
4.15	Cerea		Icosandria	Monogynia	Cactus	Caryophyllales	Cactaceae (cactus)	Selenicerus (grandiflorus)
4.15n		Arabian jasmine	Diandria	Monogynia	Nyctanthes	Lamiales	Oleaceae	Nyctanthes
4.45	Tropaeo	Garden nasturtion, greater Indian cress	Octandria	Monogynia	Tropaeolum	Brassicales	Tropaeolaceae	Tropaeolum (majus)
4.73n			Triandria	Digynia	Aira	Poales	Poaceae (true grasses)	Aira
4.73n			Triandria	Digynia	Poa	Poales	Poaceae (true grasses)	Poa
4.73n		Cow-quakes	Triandria	Digynia	Briza	Poales	Poaceae (true grasses)	Briza
4.73n		Winged spike-grass	Triandria	Digynia	Stipa	Poales	Poaceae (true grasses)	Stipa
4.73n		Rye	Triandria	Digynia	Secale	Poales	Poaceae (true grasses)	Secale
4.73n		Bread-fruit	Monoecia	Monandria	Artocarpus	Rosales	Moraceae (mulberry, fig)	Artocarpus
4.99		Beech, chestnut, chinquapin	Monoecia	Polyandria	Fagus	Fagales	Fagaceae	Fagus

(*Continued*)

(Continued)

Initial Ref. [LP]	*Primary Ref.* [LP]	*Poetic Name* [LP]	*Common Name (species)* [LP, SV]	*Class [Linnaeus]*	*Order [Linnaeus]*	*Genus or Family [Linnaeus]*	*Order [APG III System; * = not an angiosperm]*	*Family (English name) [APG III System]*	*Genus (species) [APG III System]*
4.103n			Cullions, dog's stones, fool's stones, purple bird's nest, soldier's cullions	Gynandria	Diandria	Orchis	Asparagales	Orchidaceae (orchid)	Orchis
4.103n			Saffron	Triandria	Monogynia	Crocus	Asparagales	Iridaceae	Crocus
4.103n			Conval lily, ladder to heaven, Lily of the valley, May lily, one blade, Solomon's seal	Hexandria	Monogynia	Convallaria	Asparagales	Asparagaceae	Convallaria
4.103n				Hexandria	Monogynia	Alstroemeria	Liliales	Alstroemeriaceae	Alstroemeria (ligtu)
4.114n			Devil's bit, scabious	Tetrandria	Monogynia	Scabiosa	Dipsacales	Caprifoliaceae (honeysuckle)	Scabiosa
4.121n			Candleberry myrtle (cerifera), dutch myrtle, gale (gale), sweet gale	Dioecia	Tetrandria	Myrica	Fagales	Myricaceae	Myrica (gale)
4.161n	4.270 (aegagropila), 4.366 (polymorpha)	Aega (4.270), Proteus-lover (4.366)		Cryptogamia	Algae	Conferva	Cladophorales*	Cladophoraceae	Cladophora (aegagropila)
4.170		Trapa	Water caltrops	Tetrandria	Monogynia	Trapa	Myrtales	Lythraceae	Trapa
4.170n			Drop-water, hemlock drop-wort	Pentandria	Digynia	Oenanthe	Apiales	Apiaceae (carrot, celery, parsley)	Oenanthe
4.170n			Anise, burnet saxifrage	Pentandria	Digynia	Pimpinella	Apiales	Apiaceae (carrot, celery, parsley)	Pimpinella
4.170n			Celery, celeriac, parsley, petroseline wortle (petroselinum)	Pentandria	Digynia	Apium	Apiales	Apiaceae (carrot, celery, parsley)	Apium
4.170n			Skirret, water-parsnep	Pentandria	Digynia	Sium	Apiales	Apiaceae (carrot, celery, parsley)	Sium

4.170n		Flowering rush, water gladiole	Enneandria	Hexagynia	Butomus	Alismatales	Butomaceae	Butomus
4.207	Caryo	Clove	Polyandria	Monogynia	Caryophyllus	Myrtales	Myrtaceae (myrtle)	Syzygium (aromaticum)
4.207n		Calve's-snout, female speedwell, fluellin, snap-dragon, toad flax (linaria)	Didynamia	Angiospermia	Antirrhinum	Lamiales	Scrophulariaceae	Linaria (vulgaris)
4.207n		Pellitory, pellitory of the wall	Polygamia	Monoecia	Parietaria	Rosales	Urticaceae (nettle)	Pouzolzia
4.207n		Brook lime, Paul's betony, rock germander, speedwell	Diandria	Monogynia	Veronica	Lamiales	Plantaginaceae (plantain)	Veronica
4.207n		(European) money-wort (europaea)	Didynamia	Angiospermia	Sibthorpia	Lamiales	Plantaginaceae (plantain)	Sibthorpia (europaea)
4.207n		Golden saxifrage	Decandria	Digynia	Chrysoplenium	Saxifragales	Saxifragaceae	Chrysopleni-um
4.207n		Marsh pennywort, water navel-wort, water pennywort	Pentandria	Digynia	Hydrocotyle	Apiales	Araliaceae (Aralia or ivy)	Hydrocotyle
4.237	Chunda	French honeysuckle, moving plant (movens)	Diadelphia	Decandria	Hedysarum	Fabales	Fabaceae (legume)	Hedysarum
4.237n			Cryptogamia	Algae	Marchantia	Marchantiales	Marchantiaceae	Marchantia (polymorpha)
4.259	Muschus	Coral-moss	Cryptogamia	Algae	Lichen	Lecanorales*	Cladoniaceae	Cladonia (rangiferina)
4.298	Truffelia	Puff-ball, truffle	Cryptogamia	Fungi	Lycoperdon	Agaricales*	Agaricaceae	Lycoperdon
4.328n		Apple, codlin tree, pear	Icosandria	Pentagynia	Pyrus	Rosales	Rosaceae (rose)	Pyrus
4.357	Byssus		Cryptogamia	Algae	Byssus	various*	various	various

(*Continued*)

(Continued)

Initial Ref. [LP]	*Primary Ref.* [LP]	*Poetic Name* [LP]	*Common Name (species)* [LP, SV]	*Class [Linnaeus]*	*Order [Linnaeus]*	*Genus or Family [Linnaeus]*	*Order [APG III System; * = not an angiosperm]*	*Family (English name) [APG III System]*	*Genus (species) [APG III System]*
4.366n			Bastard medic, Burgundian hay, hedge hog, horns, lucern grass, medic, moon trefoil, sea medic, snail clover, snail trefoil, snails	Diadelphia	Decandria	Medicago (polymorphia)	Fabales	Fabaceae (legume)	Medicago
			Alkali, jointed glass-wort, kelp, sal kali, salt-wort	Monandria	Monogynia	Salicornia	Caryophyllales	Amaranthaceae (amaranth)	Salicornia
4.388		Adonis	Bird's eye, fennel-leaved black hellebore, pheasant's eye	Polyandria	Polygynia	Adonis	Ranunculales	Ranunculaceae (buttercup, crowfoot)	Adonis

DESCRIPTIONS OF THE ILLUSTRATIONS

The Loves of the Plants

Illustration LP1

Frontispiece
Designed by Emma Crewe (English, c. 1768–1850)
Engraved by Samuel Alken (English, 1756–1815)

Inscription: [*1LP1789–4LP1794*] FLORA at Play with CUPID. | Emma Crewe inv.[t.] | | S. Alken fecit
[*5LP1799*] FLORA at Play with CUPID.

Imprint: [*3LP1791–4LP1794*] Published June 1st. 1791, by J Johnson, in S[t]. Pauls Church Yard, London.
[*5LP1799*] London: Published June 1st., 1799. by J. Johnson S[t]. Pauls Church Yard.

Darwin asked Alken about the possibility of "making Flora and Cupid smile" (*Letters* 89–11: 343). Alken replied that he could "not well alter the faces, but [could] retouch the other parts for 5 or 6 guineas" (*Letters* 89–12: 345). *2LP1790+* show signs of retouching in the faces, foliage and clouds.

The goddess of flowers (a "flora" is also a "descriptive catalog of the plants of any geographical area", as Linnaeus' *Flora suecica* of 1745 [*OED*]) reposes in her bower. She has exchanged roles with Cupid: she holds his bow and arrow in her lap with his quiver at her feet; dressed in a gardener's hat and apron, he bears away her rake and spade. The flowers in her hair and basket, like those that festoon the oval frame, are most likely generic rather than particular. In a note to *The Unsex'd Females*, Richard Polwhele judged Crewe to have "overstepped the modesty of nature, by giving the portrait an air of voluptuousness too luxuriously melting" (27).

Illustration LP2

Tailpiece to Proem
Version shown: *2LP1790*

[*1LP1789*] Engraved by Edward Stringer (English, 1749–1809)
[*2LP1790+*] Redesigned, engraved by Thomas Holloway (English, 1748–1827)
Inscription: [*1LP1789*] E Stringer sclp.
[*2LP1790+*] T. Holloway. sculp.

Darwin received an initial Cupid design from Susannah Wedgwood, which he rejected as "too formal a figure" (*Letters* 88–3: 302). The vignette by Stringer that accompanied *1LP1789* shows Cupid in his spiritual character, in contrast to the more lascivious one of the frontispiece; see *EV* 1.413n on the "two Cupids." He holds by the wing a butterfly representing Psyche or the soul. The marriage of Cupid and Psyche is described in *EV* 4.47–62; they also figure prominently in his analysis of the Portland Vase, *AN* 32.17. The lowered torch, a death-emblem (see *AN* 32.25), is "warming" the liberated soul (*Letters* 89–11: 343).

In the end, Darwin found Stringer's design to have been "done intollerably" and asked Josiah Wedgwood to supply an alternative, shown here, for the next edition (*Letters* 89–11: 343). In this latter, engraved by Holloway, Cupid faces right rather than left and is inflaming a flower rather than a butterfly. It appears that Darwin preferred the "*antique*" iconography of the butterfly to the "modern" one of the flower, but accepted the latter as more "elegant" in execution (*Letters* 89–12: 345).

Illustrations LP3–4

Accompany Preface 3–39
Version shown: *2LP1790*
[*1LP1789*] Engraved by Stringer
[*2LP1790–4LP1794*] Re-engraved by Frederick Nodder (English? c. 1750–c. 1800)
[*5LP1799*] Re-engraved
Inscription: [*1LP1789*] E Stringer Lich.[d]. Sclp 1788 (Illus. 3); E Stringer Lichd Sclp 1788 (Illus. 4)
[*2LP1790–4LP1794*] F.P. Nodder Sculp.[t]

The source of these figures is Pierre Bulliard (French, 1742–93), *Dictionnaire élémentaire de botanique* (Paris, 1783), pl. 2 ("Systeme sexuel de Linnaeus")—a single illustration corresponding to Darwin's two.

The versions from *1LP1789* (with the figures in Illus. 3 reversed) are two of the three "bad plates" that Darwin had Nodder re-engrave (*Letters* 89–12: 345).

Illustration LP5

Accompanies *LP* 1.61–64
Version shown: *2LP1790*
[*1LP1789*] Engraved by Stringer?
[*2LP1790–4LP1794*] Re-engraved by Nodder
[*5LP1799*] Re-engraved

Inscription: [*1LP1789*] E. Stringer Scl^{p}. | | Méadia
[*2LP1790–4LP1794*] Engrav'd by F.P. Nodder. | | Meadia
[*5LP1799*] Meadia

One of the three "bad plates" from *1LP1789* that Darwin had Nodder re-engrave (*Letters* 89–12: 345).

Illustration LP6

Accompanies *LP* 1.117–24
[*1LP1789–4LP1794*] Engraved by Nodder
[*5LP1799*] Re-engraved
Inscription: [*1LP1789–4LP1794*] Gloriosa Superba. | Engraved by Fred.^{k} P. Nodder.
[*5LP1799*] Gloriosa Superba.

Darwin told Wedgwood that "If Miss W. [Susannah] has a colour'd print of Gloriosa or of Vallisneria—or even an uncolour'd one—I should be much obliged to her for a copy of it or them" (*Letters* 88–3: 302)

Illustration LP7

Accompanies *LP* 1.131n
[*1LP1789–4LP1794*] Engraved by Nodder
[*5LP1799*] Re-engraved
Inscription: [*1LP1789–4LP1794*] Engrav'd by F.P. Nodder. | Dionæa Muscipula.
[*5LP1799*] Inscription: Dionæa Muscipula.

Darwin says that "of this plant I was favor'd with an elegant colour'd drawing by Miss Maria Jacson" (*LP* 1.131n).

Illustration LP7a *[2LP1790+]*

Accompanies insertion to *LP* 1.142; see textual notes
[*2LP1790–4LP1794*] Engraved by Nodder
[*5LP1799*] Re-engraved
Inscription: [*2LP1790–4LP1794*] Engrav'd by F.P. Nodder. | Amaryllis formosissima.
[*5LP1799*] Amaryllis formosissima.

Illustration LP7b *[5LP1799]*

Accompanies *LP* 1.247–56 and n
Inscription: The Tartarian Lamb. | Barometz.

Copied from a plate by Joseph Halfpenny (English, 1748–1811) in John Evelyn (English, 1620–1706), *Silva ... to Which Is Added the Terra*, ed. Alexander Hunter (Scottish, c. 1729–1809), 2 vols. (York, 1786), opp. 2.48.

Illustration LP8

Accompanies *LP* 1.359–72
[*1LP1789–4LP1794*] Engraved by Nodder
[*5LP1799*] Re-engraved
Inscription: [*1LP1789–4LP1794*] Vallisneria Spiralis. | Engraved by Fred.[k] P. Nodder.
[*5LP1799*] Vallisneria Spiralis.

Darwin told Wedgwood that "If Miss W. [Susannah] has a colour'd print of Gloriosa or of Vallisneria—or even an uncolour'd one—I should be much obliged to her for a copy of it or them" (*Letters* 88–3: 302)

Illustration LP8a *[5LP1799]*

Accompanies *LP* 3.51–78
Designed by Henry Fuseli, born Johann Heinrich Füssli (Swiss, 1741–1825)
Engraved by Holloway
Imprint: London. Published June 1, 1791. by J. Johnson, S[t]. Pauls Church Yard.
Inscription: H Fuseli RA Pinxit. | | T. Holloway. Sculp[t]. | Nightmare.

Fuseli executed several versions of *The Nightmare*, the first and perhaps best-known in 1781 (oil on canvas, 101.6 x 126.7 cm; now at the Detroit Institute of Arts). It is said to have been inspired by Fuseli's unrequited love for Anna Landolt (niece of the physiognomist Johann Kaspar Lavater), to whom the woman bears a resemblance, and who is believed to be represented in an unfinished portrait that appears on the reverse of the canvas. The painting was exhibited at the Royal Academy in 1782 and caused an immediate sensation, spawning a host of emulators, plagiarists and parodists. Thomas Burke's (English 1749–1815) stipple engraving based on it was published in 1783, inscribed with four lines by Darwin (closely resembling *LP* 3.51–54). A twenty-six-line poem titled "The Night Mare" (eventually reworked as *LP* 3.51–78) appeared in the *Universal Magazine of Knowledge and Pleasure* 71.494 (September 1782): 165.

The version of 1790–91 (oil on canvas, 75.5 x 64 cm; now at the Goethe-Museum, Frankfurt) differs from the previous in that "it is upright; the horse is white; the incubus is more cat-like and smiles while playing a pipe; the table [with a dressing-mirror, a perfume vial and other toiletries] is behind and not in front of the bed; the girl lies with her head to the left and her arm hanging down towards the outer edge of the canvas; she lies on a bolster, as if to ease impregnation; and her knees are raised" (Nicolas Powell, *Fuseli: The Nightmare* [New York: Viking

P, 1973], 98). Holloway's engraving of this version was issued as an octavo print in 1791 before being incorporated into *5LP1799*.

The *mare* of *nightmare* is derived from "*mara*, the name of a spirit imagined by the nations of the north to torment sleepers" (Samuel Johnson, *A Dictionary of the English Language*, 2 vols. [London, 1755–56]). This spirit was also known as a *hag*, and its nocturnal visitations as *hag-riding*. The horse may therefore be its means of transportation, as is clearly shown in Fuseli's c. 1793 painting of an incubus exiting the bedroom window of two awakening girls. *Mare* as a female horse has an altogether separate etymology; Darwin confounds the issue by saying that the "squab Fiend" "flits" "on his NIGHTMARE" (*LP* 3.51–52). The horse does not appear in Fuseli's original sketches, however, and it is certainly possible that it was added in response to Darwin's lines rather than vice-versa (Powell 59).

Illustration LP8b *[2LP1790+]*

Accompanies *LP* 4.237–40
[*2LP1790–4LP1794*] Engraved by Nodder
[*5LP1799*] Re-engraved
Inscription: [*2LP1790–4LP1794*] Engrav'd by F.P. Nodder | Hedysarum gyrans.
[*5LP1799*] Hedysarum gyrans.

Illustration LP9

Accompanies *LP* Supplement 2–3
[*1LP1789–4LP1794*] Engraved by Nodder
[*5LP1799*] Re-engraved
Inscription: [*1LP1789–4LP1794*] Drawn & Engrav'd by F.P. Nodder, Botanic Painter to her Majesty. | Apocynum androsæmifolium.
[*5LP1799*] Inscription: Apocynum androsæmifolium.

TEXTUAL NOTES TO *THE LOVES OF THE PLANTS*

PROEM

Appears in ***1LP1789***, *2LP1790*, *3LP1791*, *4LP1794*, *4EV1799*
2 leasure] leisure *4EV1799*

PREFACE

Appears in ***1LP1789***, *2LP1790*, *3LP1791*, *4LP1794*, *5LP1799*
3 Dictionaire] Dictionnaire *5LP1799*
20 No. xv] No. 15 *2LP1790*
22 the filaments] their filaments *2LP1790+*
25 as in first] as in the first *3LP1791+*
43 VEGETABLES] VEGETABLES* *3LP1791+*
44 PLANTS] PLANTS† *3LP1791*, *4LP1794* PLANTS* *5LP1799*
45 *The SYSTEM* etc.] [footnote] **The SYSTEM* etc. *3LP1791+*
45 18] **eighteen** *3LP1791+*
46 *The FAMILIES* etc.] [footnote] †*The FAMILIES* etc. *3LP1791*, *4LP1794* **The FAMILIES* etc. *5LP1799*
46 16] sixteen *3LP1791+*

CANTO 1

1.9 croud] crowd *2LP1790+*
1.10n male] males *2LP1790+*
1.51n other?] other. *4LP1794+*
1.60n pleased] delighted *3LP1791+*
1.60n prey] pray *3LP1791+*
1.65n The Opulus is ... of all things] *2LP1790+; in Supplement, with heading: Addition to the note on Curcuma. 1LP1789*
1.65n existance] *Supp. 1LP1789* existence *2LP1790+*
1.65n globe, and consonant ... of all things] *Supp. 1LP1789* globe *in text 1LP1789*

1.73n Dict. art. coniferæ] *insert after*: The gates of St. Peter's church at Rome, which had lasted from the time of Constantine to that of Pope Eugene the fourth, that is to say eleven hundred years, were of Cypress, and had in that time suffered no decay. According to Thucydides, the Athenians buried the bodies of their heroes in coffins of Cypress, as being not subject to decay. A similar durability has also been ascribed to Cedar. Thus Horace,

———*speramus carmina fingi*
Posse linenda cedro, & lævi servanda cupresso. 2LP1790+

1.86n There is an instance ... Bladh. Amæn. Acad. V. 7.] *2LP1790+; in Supplement, with heading: Addition to the note on Anthoxanthum. 1LP1789*

1.119n stand up] *2LP1790+* hang down *errata 1LP1789*

1.119n have numerous stamens] *insert after*: In the Kalmia the ten stamens lie round the pistil like the radii of a wheel; and each anther is concealed in a nich of the corol to protect it from cold and moisture; these anthers rise separately from their niches, and approach the pistil for a time, and then recede to their former situations. *2LP1790+*

1.125–30 So Ninon pruned ... from his heart!]

So, in her wane of beauty, Ninon won
With fatal smiles her gay unconscious son.—
Clasp'd in his arms she own'd a mother's name,—
"Desist, rash youth! restrain your impious flame,
"First on the bed your infant-form was press'd,
"Born by my throes, and nurtured at my breast."—
Back as from death he sprung, with wild amaze
Fierce on the fair he fix'd his ardent gaze;
Dropp'd on one knee, his frantic arms outspread,
And stole a guilty glance toward the bed;
Then breath'd from quivering lips a whisper'd vow,
And bent on heaven his pale repentant brow;
"Thus, thus!" he cried, and plung'd the furious dart,
And live and love gushed mingled from his heart. *2LP1790+*

1.131n See Cerea] *insert after*: See additional notes at the end of the poem. *2LP1790+*

1.138 Sirens] Syrens *2LP1790+*

1.139 Limed with their art] Limed by their art *2LP1790+*

1.142 the honey purchas'd with your lives!] *insert after*:

With heaven's high vault condensing clouds deform,
Fair Amaryllis flies the incumbent storm,
Seeks with unsteady step the shelter'd vale,

And turns her blushing beauties from the gale.—
Six rival youths, with soft concern impres'd,
Calm all her fears, and charm her cares to rest.—
So shines at eve the sun-illumin'd fane,
Lifts its bright cross, and waves its golden vane;
From every breeze the polish'd axle turns,
And high in air the dancing meteor burns. *2LP1790+*

1.142a2 the incumbent storm] *insert n: Amaryllis*. ... Formosissima. Most beautiful Amaryllis. Six males, one female. Some of the bell-flowers close their apertures at night, or in rainy or cold weather, as the convolvulus, and thus protect their included stamens and pistils. Other bell-flowers hang their apertures downwards, as many of the lilies; in those the pistil, when at maturity, is longer than the stamens; and by this pendant attitude of the bell, when the anthers burst, their dust falls on the stigma: and these are at the same time sheltered as with an umbrella from rain and dews. But, as a free exposure to the air is necessary for their fecundation, the style and filaments in many of these flowers continue to grow longer after the bell is open, and hang down below its rim. In others, as in the martagon, the bell is deeply divided, and the divisions are reflected upwards, that they may not prevent the access of air, and at the same time afford some shelter from perpendicular rain or dew. Other bell-flowers, as the hemerocallis and amaryllis, have their bells nodding only, as it were, or hanging obliquely toward the horizon; which, as their stems are slender, turn like a weathercock from the wind; and thus very effectually preserve their inclosed stamens and anthers from the rain and cold. Many of these flowers, both before and after their season of fecundation, erect their heads perpendicular to the horizon, like the Meadia, which cannot be explained from meer mechanism.

The Amaryllis formosissima is a flower of the last mentioned kind, and affords an agreeable example of *art* in the vegetable economy. 1. The pistil is of great length compared with the stamens; and this I suppose to have been the most unchangeable part of the flower, as in Meadia, which see. 2. To counteract this circumstance, the pistil and stamens are made to decline downwards, that the prolific dust might fall from anthers on the stigma. 3. To produce this effect, and to secure it when produced, the corol is lacerated, contrary to what occurs in other flowers of this genus, and the lowest division with the two next lowest ones are wrapped closely over the style and filaments, binding them forcibly down lower toward the horizon than the usual inclination of the bell in this genus, and thus constitutes a most elegant flower. There is another contrivance for this purpose in the Hemerocallis flava: the long pistil often is bend somewhat like the capital letter N, with design to shorten it, and thus to bring the stigma amongst the anthers. *2LP1790+*

1.143n The numerous clumps ... it is called] *2LP1790+; in Additional Notes, with tag Lead the lone pilgrim. 1LP1789*

1.143n it is called] *insert after:* There is a fossile elastic bitumen found at Matlock in Derbyshire, which much resembles these substances in its elasticity and inflammability. The thorns of the mimosa cornigere resemble cow's horns in appearance as well as in use. System of Vegetables, p. 782 *2LP1790+*

1.156 wandring] wand'ring *2LP1790* wandering *3LP1791+*

1.156 the wandring maid] *insert after*:

So WRIGHT's bold pencil from Vesuvio's hight
Hurls his red lavas to the troubled night;
From Calpè starts the intolerable flash,
Skies burst in flames, and blazing oceans dash;—
Or bids in sweet repose his shades recede,
Winds the still vale, and slopes the velvet mead;
On the pale stream expiring Zephyrs sink,
And Moonlight sleeps upon its hoary brink. *2LP1790+*

1.156a2 the troubled night] *insert n*: *Hurls his red lavas. ...* Alluding to the grand paintings of the eruptions of Vesuvius, and of the destruction of the Spanish vessels before Gibraltar; and to the beautiful landscapes and moonlight scenes, by Mr. Wright of Derby. *2LP1790+*

1.158 of her wide domains] of Orixa's plains *2LP1790+*

1.164 puny] trembling *2LP1790+*

1.165 bound her jutting breast] shook her plumy crest *2LP1790+*

1.166 In rigid mail, and shook her plumy crest] And bound in rigid mail her jutting breast *2LP1790+*

1.177n By cautiously cutting through] By cautiously cutting in the early spring through *2LP1790* By cautiously cutting in winter through *3LP1791+*

1.177n the whole flower of the next year's tulip] the whole flower of the next summer's tulip *2LP1790+*

1.177n In the seeds ... its parts. ibid.] *2LP1790+; in Supplement, with heading: Addition to the note on Tulipa. 1LP1789*

1.181–84 So to his ... thy roseate form]

So, six cold moons, the Dormouse charm'd to rest,
Indulgent Sleep! beneath thy eider breast,
In fields of Fancy climbs the kernel'd groves,
Or shares the golden harvest with his loves.—
But [Then *errata 3LP1791+*] bright from earth amid the troubled air [sky *emendation 3LP1791+*]
Ascends fair Colchica with radiant hair, [eye, *4LP1794+*] *2LP1790+*

1.190 And drives o'er ... his glittering car]

And drives on Night's blue arch his glittering car;
Hangs o'er the billowy clouds his lucid form,
Wades through the mist, and dances in the storm. *2LP1790+*

1.192 Dervice] Dervise *2LP1790+*

1.199 dewy vale, fair] marsh, imperial *2LP1790+*

1.199n and destroyed it] *insert after*: M. Brouffonet, in the Mem. de l'Acad. de Sciences for the year 1784. p. 615. after having described the motion of the Dionæ, adds, that a similar appearance has been observed in the leaves of two species of Drosera. *2LP1790+*

1.200 Her moss-wove banks, and rushy-fringed] Rush-fringed banks, and moss-embroider'd *2LP1790+*

1.211 treads] prints *2LP1790+*

1.211n and Conferva Polymorpha] and Conferva Polymorpha, and on Epidendrum *5LP1799*

1.220 her eagle nest] *insert after*:

> Her pendant eyry icy caves surround,
> Where erst Volcanos [Volcanoes *5LP1799*] min'd [mined *errata 3LP1791+*] the rocky ground. *2LP1790+*

1.220n as cabbage, turnep] *insert after*: There was formerly a Volcano on the Peake of Tenerif, which became extinct about the year 1684. Philos. Trans. In many excavations of the mountain, much below the summit, there is now found abundance of ice at all seasons. Tench's Expedition to Botany Bay, p. 12. Are these congelations in consequence of the daily solution of the hoar-frost, which is produced on the summit during the night? *2LP1790+*

1.225 Stay] Oh, stay *3LP1791+*

1.226 eagle-flight] angel-flight *3LP1791+*

1.226n apple-trees annually] *insert after*: See Epidendrum. *5LP1799*

1.245 pansied] pancied *3LP1791+*

1.249 fan'd] fann'd *2LP1790+*

1.288n The increase of ... V. I. p. 19] *2LP1790+*; *in Supplement, with heading: Addition to the note on The Swallow. 1LP1789*

1.288n tho'] *Supp. 1LP1789* though *2LP1790+*

1.294 In death-like slumbers seals] Who seals in death-like sleep *2LP1790+*

1.316 'mid the airy ocean dwells] drinks the aerial solitude *3LP1791+*

1.318 her] *2LP1790+* his *errata 1LP1789*

1.333n Zeyl. 42. 17] *insert after*: Silphium perfoliatum has a cup round every joint to reserve water after rain. It rises during the summer twelve or fourteen feet high on a slender stem, which is square, and thus is stronger to resist the winds then if it had been made round with the same quantity of materials.

The most curious plant of this kind is the Sarracenia purpurea, which resembles the Nymphœa, an aquatic plant, but catches so much water in its sessile cup-like leaves, as to enable it to live on land, a wonderful provision of nature! System. Plant. a Reichard. Vol. II. p. 577. *5LP1799*

1.341 blends] *3LP1791+* blends, *1LP1789 2LP1790*

1.341n To insects and many smaller animals their colours] The colours of insects and many smaller animals *3LP1791+*

1.353n *Pleased on the ...* advances of age] *2LP1790+; in Additional Notes 1LP1789*

1.353n Aeson's] *Add. Notes 1LP1789* Æson *2LP1790+*

1.353n advances of age] *Add. Notes 1LP1789* advances of old age *2LP1790+*

1.358 locks luxuriant] darker tresses *2LP1790+*

1.358 round his brow] *insert after*:

Where Java's isle, horizon'd with the floods,
Lifts to the skies her canopy of woods;
Pleased Epidendra climbs the waving pines,
And high in heaven the intrepid beauty shines,
Gives to the tropic breeze her radiant hair,
Drinks the blight flower, and feeds upon the air.
Her brood delighted stretch their callow wings,
As poised aloft their pendent cradle swings,
Eye the warm sun, the spicy zephyr breathe,
And gaze unenvious on the world beneath. *5LP1799*

1.358a3 the waving pines] *insert n*: *Epidendrum flos aeris...* . Of the class gynandria, or feminine males. This parasite plant is found in Java, and is said to live on air without taking root in the trees on which it grows; and its flowers resemble spiders. Syst. Veg. a Reichard. Vol. IV. p. 35. By this curious similitude the bees and butterflies are supposed to be deterred from plundering the nectaries. See Visea. *5LP1799*

1.361n as ants] as ants, Cocchus, Lampyris, Phalæna, Brumata, Lichanella *2LP1790+*

1.373 the intrepid] intrepid *2LP1790+*

INTERLUDE [1]

1.14 of poetic ornament] *2LP1790+* of poetry ornament *1LP1789*

CANTO 2

2.37 brow] brows *5LP1799*

2.38 with mingled cries they vow] they breathe their mingled vows *5LP1799*

2.52 dusky] crystal *3LP1791+*

2.58 Through] Thro' *3LP1791+*

2.60n Virg. Georg. l. I. 34] *insert after*: A new star appeared in Cassiope's chair in 1572. Herschel's Construction of the Heavens. Phil. Trans. V. 75. p. 266. *2LP1790+*

2.67n portigal] Portugal *3LP1791+*

2.93 emerging Naiads cull] *insert n*:

Emerging Naiads... . —eam circum Milesia vellera Nymphæ
Carpebant, hyali faturo fucata colore. Vir. Georg. IV. 334. *3LP1791+*;
in Additional Notes with tag: Additional note on Gossypium. 2LP1790

2.119–120 with fond officious care ... assist the fair]

her soft attention share,
The fond disciples of the studious Fair,
Hear her sweet voice, the golden process prove;
Gaze, as they learn; and, as they listen, love. *2LP1790+*

2.147 From Earth's deep chambers tore her golden stores] His treasur'd gold from Earth's deep chambers tore *2LP1790+*

2.148 ores] ore *2LP1790+*

2.208 brow] face *2LP1790+*

2.210 the east was] Assyria *3LP1791+*

2.211 To dwell] Sojourn'd *3LP1791+*

2.212 By Heaven's just vengeance changed] Changed by avenging Heaven *3LP1791+*

2.240 wave] waves *2LP1790+*

2.246 babling] babbling *2LP1790+*

2.277 Fill'd] Flush'd *2LP1790+*

2.278 as they walk] *insert after:*

With deeper sobs reviving lovers breathe,
Fair bosoms rise, and soft hearts pant beneath; *2LP1790+*

2.280 brighter] kindling *2LP1790+*

2.336 Fantastic] Suspended *4LP1794+*

2.364 his marble feet] black hoof *emendation 3LP1791+*

2.431n extream tenderness. (Geograph... . Lond.) and though] extream tenderness; and though *2LP1790+*

2.435 Benevolence] Philanthropy *2LP1790+*

INTERLUDE [2]

2.12 of the sufferers] *insert after*: An ingenious writer, who has criticised this dialogue in the English Review for August, 1789, adds, that one great source of our pleasure from scenical distress arises from our, at the time, generally contemplating one of the noblest objects of nature, that of Virtue triumphant over every difficulty and oppression, or supporting its votary under every suffering: or, where this does not occur, that our minds are relieved by the justice of some signal punishment awaiting the delinquent. *3LP1791+*

2.12 But at the exhibition] But, besides this, at the exhibition *2LP1790+*

2.13 third Canto] *2LP1790+* second Canto *errata 1LP1789*

CANTO 3

3.7n whiped] whipped *2LP1790+*

3.7n And in the very last year] And in this very year *3LP1791+*

3.74n thrist] thirst *2LP1790+*

3.76 ballances] balances *2LP1790+*

3.76 bloated] *2LP1790+* boated *errata 1LP1789*

3.170 poynards] poniards *2LP1790+*

3.179 isles] *2LP1790+* isle *errata 1LP1789*

3.184n One of the ... Printemp. p. 109] *2LP1790+; in Supplement, with heading: Addition to the note on Dictamnus. 1LP1789*

3.184n injured and destroyed] *Supp. 1LP1789* injured *2LP1790+*

3.184n Printemp. p. 109] *Supp. 1LP1789 insert after*: I sprinkled some oil of turpentine, by means of a brush, on some branches of a nectarine-tree, which was covered with the aphis; but it killed both the insect and the branches: a solution of arsenic much diluted did the same. *2LP1790+*

3.193–94 Lobelia with contagious ... light, and wings]

Lobelia's suffocating breath
Loads the dank pinion of *2LP1790+*

3.211 foamy] foaming *3LP1791+*

3.223 breathes the breeze] zephyrs blow *2LP1790+*

3.238n Iava] Java *2LP1790+*

3.238n Foereh] Foersch *5LP1799*

3.251 infant Dæmons] scion-demons *2LP1790+*

3.252 soft] faint *2LP1790+*

3.252 tender] shriller *2LP1790+*

3.258 of domestic joys] *insert after*:

With blushes bright as morn fair Orchis charms,
And lulls her infant in her fondling arms;
Soft plays *Affection* round her bosom's throne,
And guards his life, forgetful of her own.
So wings the wounded Deer her headlong flight,
Pierced by some ambush'd archer of the night,
Shoots to the woodlands with her bounding fawn,
And drops of blood bedew the conscious lawn;
There hid in shades she shuns the cheerful day,
Hangs o'er her young, and weeps her life away.

So stood Eliza on the wood-crowned height,
O'er Minden's plain, spectatress of the fight,
Sought with bold eye amid the bloody strife
Her dearer self, the partner of her life;
From hill to hill the rushing host pursued,
And view'd his banner, or believed she view'd.
Pleased with the distant roar, with quicker tread
Fast by his hand one lisping boy she led;
And one fair girl amid the loud alarm
Slept on her kerchief, cradled by her arm;
While round her brows bright beams of Honour dart,
And Love's warm eddies circle round her heart.

—Near and more near the intrepid Beauty press'd,
Saw through the driving smoke his dancing crest,
[Saw on his helm, her virgin-hands inwove,
Bright stars of gold, and mystic knots of love; *3LP1791+; in Additional Notes, with tag: Addition to Orchis. 2LP1790*]
Heard the exulting shout, "they run! they run!"
"Great God!" she cried, "He's safe! the battle's won!"
—A ball now hisses through the airy tides,
(Some Fury wing'd it, and some Demon guides!)
Parts the fine locks, her graceful head that deck,
Wounds her fair ear, and sings into her neck;
The red stream, issuing from her azure veins,
Dyes her white veil, her ivory bosom stains.—
—"Ah me!" she cried, and, sinking on the ground,
Kiss'd her dear babes, regardless of the wound;
"Oh, cease not yet to beat, thou Vital Urn!
Wait, gushing Life, oh, wait my Love's return!—
Hoarse barks the wolf, the vulture screams from far!
The angel, Pity, shuns the walks of war!—
Oh, spare ye War-hounds, spare their tender age!—
On me, on me," she cried, "exhaust your rage!"—
Then with weak arms her weeping babes caress'd,
And sighing hid them in her blood-stain'd vest.

From tent to tent the impatient warrior flies,
Fear in his heart, and frenzy in his eyes;
Eliza's name along the camp he calls,
Eliza echoes through the canvas walls;
Quick through the murmuring gloom his footsteps tread,
O'er groaning heaps, the dying and the dead,
Vault o'er the plain, and in the tangled wood,

Lo! dead Eliza weltering in her blood!—
—Soon hears his listening son the welcome sounds,
With open arms and sparkling eyes he bounds:—
"Speak low," he cries, and gives his little hand,
"Eliza sleeps upon the dew-cold sand;
Poor weeping Babe with bloody fingers press'd,
And tried with pouting lips her milkless breast;
Alas! we both with cold and hunger quake—
Why do you weep?—Mama will soon awake."
—"She'll wake no more!" the hopeless mourner cried,
Upturn'd his eyes, and clasp'd his hands, and sigh'd;
Stretch'd on the ground awhile entranc'd he lay,
And press'd his warm kisses on the lifeless clay;
And then unsprung with wild convulsive start,
And all the Father kindled in his heart;
"Oh Heaven!" he cried, "my first rash vow forgive!
These bind to earth, for these I pray to live!"—
Round his chill babes he wrapp'd his crimson vest,
And clasp'd them sobbing to his aching breast. *2LP1790+*

3.258a1 fair ORCHIS charms] *insert n*: *Orchis...* . The Orchis morio in the circumstance of the parent-root shrivelling up and dying, as the young one increases, is not only analagous to other tuberous or knobby roots, but also to some bulbous roots, as the tulip. The manner of the production of herbaceous plants from their various perennial roots, seems to want further investigation, as their analogy is not yet clearly established. The caudex, or true root, in the orchis lies above the knob; and from this part the fibrous roots and the new knob are produced. In the tulip the caudex lies below the bulb; from whence proceed the fibrous roots and the new bulbs; [and I suspect the tulip-root *2LP1790* the root *3LP1791+*], after it has flowered, dies like the orchis root; for the stem of the last year's tulip lies on the outside, and not in the center of the new bulb; which I am informed does not happen in the three or four first years when raised from seed, when it only produces a stem, and slender leaves without flowering. In the tulip-root, dissected in the early spring, just before it begins to shoot, a perfect flower is seen in its center; and between the first and second coat the large next year's bulb is, I believe, produced; between the second and third coat, and between this and the fourth coat, and perhaps further, other less and less bulbs are visible, all adjoining to the caudex at the bottom of the mother-bulb; and which, I am told, require as many years before they will flower, as the number of the coats with which they are covered. This annual reproduction of the tulip-root induces some florists to believe that tulip-roots never die naturally, as they lose so few of them; whereas the hyacinth-roots, I am informed, will not last above five or seven years after they have flowered.

The hyacinth-root differs from the tulip root, as the stem of the last year's flower is always found in the center of the root, and the new off-sets arise from the caudex below the bulb, but not beneath any of the concentric coats of the root, except the external one; hence Mr. Eaton, an ingenious florist of Derby, to whom I am indebted for most of the observations in this note, concludes, that the hyacinth-root does not perish annually after it has flowered like the tulip. Mr. Eaton gave me a tulip root which had been set too deep in the earth, and the caudex had elongated itself near an inch, and the new bulb was formed above the old one, and detached from it, instead of adhering to its side. [See addit. Notes to Vol. I No. XIV. *3LP1791+*]

The caudex of the ranunculus, cultivated by the florists, lies above the claw-like root; in this the old root or claws die annually, like the tulip and orchis, and the new claws, which are seen above the old ones, draw down the caudex lower into the earth. The same is said to happen to Scabiosa, or Devils bit, and some other plants as valerian and greater plaintain; the new fibrous roots rising round the caudex above the old ones, the inferior end of the root becomes stumped, as if cut off, after the old fibres are decayed, and the caudex is drawn into the earth by these new roots. See Arum and Tulipa. *2LP1790+*

3.259n not been investigated] *insert after*: Other plants are furnished with tendrils for the purpose of climbing: if the tendril meets with nothing to lay hold of in its first revolution, it makes another revolution; and so on till it wraps itself quite up like a cork-screw; hence, to a careless observer, it appears to move gradually backwards and forwards, being seen sometimes pointing eastward and sometimes westward. *2LP1790+*

3.301 Prometheus, fearless of his] Prometheus braved the Thunderer's *2LP1790+*

3.302 the throne of Jove forbidden] his blazing throne etherial [ethereal *4LP1794+*] *2LP1790+*

3.309 regardless of] exulting in *2LP1790+*

3.343n *Rolls his brineless* ... ocean perpetually passes] *2LP1790+; in Additional Notes, with tag: Rolls his brineless tide. 1LP1789*

3.345n Sir Hans Sloane ... of the sea] *2LP1790+; in Supplement, with heading: Addition to the note on Cassia. 1LP1789*

3.345n Philos. Trans. abridged. V. 3] *Supp. 1LP1789* Phil. Trans. abridged, Vol. III *2LP1790+*

3.345n subsiding of the sea] subsiding of the sea, which might probably lay all those islands into one, or join them to the continent *3LP1791+*

3.381–82 The close recesses ... arbiter of Sin]

Throned in the vaulted heart, his dread resort,
Inexorable Conscience holds his court *2LP1790+*

3.383 plotts] plots *2LP1790+*

3.384 Lights his dark mind] Bares his mask'd brow *2LP1790+*

3.385 wrap'd] wrapp'd *2LP1790+*

3.392 vernal suns, that gild the rising] rising suns that gild the vernal *2LP1790+*

3.393 breaks] flows *4LP1794+*

3.394 For other's woe down Virtue's manly cheeks] Down Virtue's manly cheek for others' woes *4LP1794+*

INTERLUDE [3]

3.52 beauty of the versification] *insert after*: The following line translated from Juvenal by Dr. Johnson, is much superior to the original:

Slow rises Worth by Poverty depress'd.

The original is as follows:

Difficile emergunt, quorum virtutibus obstat,
Res angusta domi. *3LP1791+*

3.55 Dr. Yonge's] Dr. Young's *2LP1790+*

3.55 exotic] *2LP1790+* exoitic *errata 1LP1789*

CANTO 4

4.15n labdiniferus ... labdaniferus] labdiniferus ... lubdiniferus *2LP1790* labdaniferus ... labdaniferus *3LP1791+*

4.35n F. N. Mundy] F.N.C. Mundy *3LP1791+*

4.35n leaving Needwood-Forest] *insert after*: See the passage in the notes at the end of this volume. *3LP1791+*

4.45n The nectary of ... from the corol] The ceasing to shine of this plant after twilight might induce one to conceive, that it absorbed and emitted light, like the Bolognian Phosphorus, or calcined oyster-shells, so well explained by Mr. B. Wilson, and by T.B. Beccari. Exper. on Phosphori, by B. Wilson. Dodsley. The light of the evening, at the same distance from noon, is much greater, as I have repeatedly observed, than the light of the morning: this is owing, I suppose, to the phosphorescent quality of almost all bodies, in a greater or less degree, which thus absorb light during the sun-shine, and continue to emit it again for some time afterwards, though not in such quantity as to produce apparent scintillations. The nectary of this plant grows from what is supposed to be the calyx; but this supposed calyx is coloured; and perhaps, from this circumstance of its bearing the nectary, should rather be esteemed a part of the coral [corol *LP41794+*]. See an additional note at the end of the poem *2LP1790+*

4.51n *So shines the* ... burnt to death] *2LP1790+; in Additional Notes, with tag: So shines the glow-fly. 1LP1789*

4.70 calm] slow *2LP1790+*

4.73n *Ovena*] *Avena 5LP1799*

4.102 the favour'd name] *insert after*:

Green swells the beech, the widening knots improve,
So spread the tender growths of culture'd [living *emendation 3LP1791+*] love;
Wave follows wave, the letter'd lines decay,
So Love's soft forms neglected [uncultured *emendation 3LP1791+*] melt away. *3LP1791+; in Additional Notes, with tag:* The following lines were by mistake omitted; they were designed to have been inserted after l. 102 *2LP1790+*

From Time's remotest dawn where China brings
In proud succession all her Patriot-Kings;
O'er desert-sands, deep gulfs, and hills sublime,
Extends her massy wall from clime to clime;
With bells and dragons crest her Pagod-bowers,
Her silken palaces, and porcelain towers;
With long canals a thousand nations laves;
Plants all her wilds, and peoples all her waves;
Slow treads fair CANNABIS the breezy strand,
The distaff streams dishevell'd in her hand;
Now to the left her ivory neck inclines,
And leads in Paphian curves its azure lines;
Dark waves the fringed lid, the warm cheek glows,
And the fair ear the parting locks disclose;
Now to the right with airy sweep she bends,
Quick join the threads, the dancing pole depends.
—*Five* Swains attracted guard the Nymph, by turns
Her grace inchants them, and her beauty burns;
To each She bows with sweet assuasive smile,
Hears his soft vows, and turns her spole the while.

So when with light and shade, concordant strife!
Stern CLOTHO weaves the chequer'd thread of life;
Hour after hour the growing line extends,
The cradle and the coffin bound its ends;
Soft cords of silk the whirling spoles reveal,
If smiling Fortune turn the giddy wheel;
But if sweet Love with baby-fingers twines,
And wets with dewy lips the lengthening lines,
Skein after skein celestial tints unfold,
And all the silken tissue shines with gold. *2LP1790+*

4.102a9 the breezy strand] *insert n*: *Cannabis...* . Chinese Hemp. Two houses. Five males. A new species of hemp, of which an account is given by K. Fitzgerald, Esq. in a letter to Sir Joseph Banks, and which is believed to be much superior to the hemp of other countries. A few seeds of this plant were sown in England on the 4th of June, and grew to fourteen feet seven inches in height by the middle of October: they were nearly seven inches in circumference, and bore many lateral branches, and produced very white and touch fibres. At some parts of the time these plants grew nearly eleven inches in a week. Philos. Trans. Vol. LXXII. p. 46. *2LP1790+*

4.102a12] *insert n*: *Paphian curves...* . In his ingenious work, entitled, The Analysis of Beauty, Mr. Hogarth believes that the triangular glass, which was dedicated to Venus in her temple at Paphos, contained in it a line bending spirally round a cone with a certain degree of curviture; and that this pyramidical outline and serpentine curve constitute the principles of Grace and Beauty. *2LP1790+*

4.103n as good salep] *insert after*: The roots of the Hyacinth, I am informed, are equally insipid, and might be used as an article of food.

4.103n producing many bulbs] *insert after*: Vol. VI. p. 120. *2LP1790+*

4.121n *Cape, V. I. p. 345*] *Cape*, V. I. 345 *2LP1790*

4.121n *Cape, V. I. p. 345*] *insert after*: Du Halde [Du Valde *5LP1799*] gives an account of a white wax [*2LP1790 4LP1794+* white-wax *3LP1791*] made by small insects round the branches of a tree in China in great quantity, which is there collected for economical and medical [*2LP1790* medical and economical *3LP1791+*] purposes. The tree is called Tong-tsin. Description [*2LP1790* Descript. *3LP1791 5LP1799* Descrip. *4LP1794*] of China, Vol. I. p. 230. *3LP1791+; in Additional Notes, with tag: Additional note to Bellis. 2LP1790+*

4.135 wing, above,] pinion twine *2LP1790+*

4.136 From bough to bough extend] Or stretch from bough to bough *2LP1790+*

4.137 Scar] Scare *2LP1790+*

4.170n a respiratory one] a respiratory one; and probably the lamprey, so frequent in the month of April both in the Severn and Derwent, inspires and expires water on the seven holes on each side of the neck, which thus performs the office of the gills of other fish *5LP1799*

4.186 ever-changing dyes] *insert after*:

Where Andes, crested with volcanic beams,
Sheds a long line of light on Plata's streams;
Opes all his springs, unlocks his golden caves,
And feeds and freights the immeasurable waves;
Delighted OCYMA at twilight hours
Calls her light car, and leaves the sultry bowers;—
Love's rising ray, and Youth's seductive dye,
Bloom'd on her check, and brighten'd in her eye;
Chaste, pure, and white, a zone of silver graced
Her tender breast, as white, as pure, as chaste;—

By *four* fond swains in playful circles drawn,
On glowing wheels she tracks the moon-bright lawn,
Mounts the rude cliff, unveils her blushing charms,
And calls the panting zephyrs to her arms.
Emerged from ocean springs the vaporous air,
Bathes her light limbs, uncurls her amber hair,
Incrusts her beamy form with films saline,
And Beauty blazes through the crystal shrine.—
So with pellucid studs the ice-flower gems
Her rimy foliage, and her candied stems.
So from his glassy horns, and pearly eyes,
The diamond-beetle darts a thousand dyes;
Mounts with enamel'd wings the vesper gale,
And wheeling shines in adamantine mail.

Thus when loud thunders o'er Gomorrah burst,
And heaving earthquakes shook his realms accurst,
An Angel-guest led forth the trembling Fair
With shadowy hand, and warn'd the guiltless pair;
"Haste from these lands of sin, ye Righteous! fly,
"Speed the quick step, nor turn the lingering eye!"—
—Such the command, as fabling Bards indite,
When Orpheus charm'd the grisly King of Night;
Sooth'd the pale phantoms with his plaintive lay,
And led the fair Assurgent into day.—
Wide yawn'd the earth, the fiery tempest flash'd,
And towns and towers in one vast ruin crash'd;—
Onward they move,—loud horror roars behind,
And shrieks of Anguish bellow in the wind.
With many a sob, amid a thousand fears,
The beauteous wanderer pours her gushing tears;
Each soft connection rends her troubled breast,
—She turns, unconscious of the stern behest!—
"I faint! I freeze! just Heaven regards my fault,
"Numbs my cold limbs, and hardens into salt!—
"Not yet, not yet, your dying Love resign!—
"This last, last kiss receive!—no longer thine!"—
She said, and ceased,—her stiffen'd form He press'd,
And strain'd the briny column to his breast;
Printed with quivering lips the lifeless snow,
And wept, and gazed the monument of woe.—
So when Æneas through the flames of Troy
Bore his pale fire, and led his lovely boy;
With loitering step the fair Creusa stay'd,

And Death involved her in eternal shade.—
—Oft the lone Pilgrim, that his road forsakes,
Marks the wide ruins, and the sulphur'd lakes;
On mouldering piles amid asphaltic mud
Hears the hoarse bittern, where Gomorrah stood;
Recalls the unhappy Pair with lifted eye,
Leans on the crystal tomb, and breathes the silent sigh. *2LP1790+*

4.186a5 at twilight hours] *insert n*: *Ocymum salinum*. l. 221. Saline Basil. Class Two Powers. The Abbé Molina, in his History of Chili, translated from the Italian by Abbé Grewvel, mentions a species of Basil, which he calls Ocymum salinum: he says it resembles the common basil, except that the stalk is round and jointed; and that though it grows 60 miles from the sea, yet every morning it is covered with saline globules, which are hard and splendid, appearing at a distance like dew; and that each plant furnishes about half an ounce of fine salt every day, which the peasants collect, and use as common salt, but esteem it superior in flavour.

As an article of diet, salt seems to act simply as a stimulus, not containing any nourishment, and is the only fossil substance which the caprice of mankind has yet taken into their stomachs along with their food; and, like all other unnatural stimuli, is not necessary to people in health, and contributes to weaken our system; though it may be useful as a medicine. It seems to be the immediate cause of sea scurvy, as those patients quickly recover by the use of such provisions; and is probably a remote cause of scrophula (which consists in the want of irritability in the absorbent vessels), and is therefore serviceable to these patients; as wine is necessary to those whose stomachs have been weakened by its use. The universality of the use of salt with our food, and in our cookery, has rendered it difficult to prove the truth of these observations. I suspect that flesh-meat cut into thin slices, either raw or boiled, might be preserved in coarse sugar or treacle; and thus a very nourishing and salutary diet might be presented to our seamen. See note on Salt-rocks, in Vol. I. Canto II. If a person unaccustomed to much salt should eat a couple of red-herrings, his insensible perspiration will be so much increased by the stimulus of the salt, that he will find it necessary in about two hours to drink a quart of water: the effects of a continued use of salt in weakening the action of the lymphatic system may hence be deduced. *2LP1790+*

4.186a19 ice-flower gems] *insert n: Ice-flower...* . Mesembryanthemum crystallinum. *2LP1790+*

4.189n See Rubia] *insert after*: When the petals of the tulip become tipped with many colours, the plant loses almost half of its height; and the method of making them thus break into colours is by transplanting them into a meagre or sandy soil, *after they have previously enjoyed a richer soil*: hence it appears, that the plant is weakened when the flower becomes variegated. See note on Anemone. For the acquired habits of vegetables, see Tulipa, Orchis.

The roots of the Arum are scratched up and eaten by thrushes in severe snowy seasons. White's Hist. of Selbourn, p. 43. *2LP1790+*

4.199 Fits] Plaits *2LP1790+*

4.205 pards] bards *2LP1790 3LP1791*

4.207n Pulteney's View of Linneus, p. 250] Pulteney's View of Linneus, p. 253

4.230n *Phil. Transact. V.* 71] *Phil. Transact.* [*Trans. 3LP1791+*] *V.* [Vol. *5LP1799*] *LXXI.*

The Rev. Mr. Sterling gives an account of a darkness for six or eight hours at Detroit in America, on the 19th of October, 1762, in which the sun appeared as red as blood, and thrice its usual size: some rain falling, covered white paper with dark drops, like sulphur or dirt, which burnt like wet gunpowder, and the air had a very sulphureous smell. He supposes this to have been emitted from some distant earthquake or volcano. Philos. Trans. V. LIII. p. 63 *2LP1790+*

4.237n movens] gyrans *2LP1790+*

4.237n to animal life] *insert after*: A more particular account, with a good print of the Hedysarum gyrans is given by M. Broussonet in a paper on vegetable motions in the Histoire de l'Academie des Sciences. Ann. 1784, p. 609. *2LP1790+*

4.244 silky] gauzy *2LP1790+*

4.247 to hide, betrays] *insert after*:

Cold from a thousand rocks, where Ganges leads
The gushing waters to his sultry meads;
By moon-crown'd mosques with gay reflections glides,
And vast pagodas trembling on his sides;
With sweet loquacity Nelumbo sails,
Shouts to his shores, and parleys with his gales;
Invokes his echoes, as she moves along,
And thrills his ripling surges with her song.
—As round the Nymph her listening lovers play,
And guard the Beauty on her watery way;
Charm'd on the brink relenting tygers gaze,
And pausing buffaloes forget to graze;
Admiring elephants forsake their woods,
Stretch their wide ears, and wade into the floods;
In silent herds the wondering sea-calves lave,
Or nod their slimy foreheads o'er the wave;
Poised on still wing attentive vultures sweep,
And winking crocodiles are lull'd to sleep. *4LP1794+*

4.247a5 loquacity Nelumbo sails] *insert n*: *Nelumbo...* . Nymphæa Nelumbo. A beautiful rose-red flower on a receptacle as large as an artichoke. The capsule is perforated with holes at the top, and the seeds rattle in it. Perfect leaves are seen in the seeds before they germinate. Linneus, who has enlisted all our senses into the service of botany, has observed this rattling of the Nelumbo; and mentions what he calls an electric murmur, like distant thunder in hop-yards, when the wind blows, and asks the cause of it. We have one kind of pedicularis in our meadows, which has obtained the name of rattle-grass, from the rattling of its dry seed-vessels under our feet. *4LP1794+*

4.259n See notes on part I] See note on Anemone *2LP1790+*
4.297 unfathom'd cells] their shadowy ailes *2LP1790+*
4.298 dwells] smiles *2LP1790+*
4.307–26 O'er the gay ... the mingled blaze]

In glittering threads along reflective walls
The warm rill murmuring twinkles, as it falls;
Now sink the Eolian strings, and now they swell,
And Echoes woo in every vaulted cell;
While on white wings delighted Cupids play,
Shake their bright lamps, and shed celestial day. *2LP1790+*

4.311n *Scoop'd in the briny* ... of these observations] *om. 2LP1790+*
4.353n Keir in the Philos. Trans. Vol.] Keir, in the Philos. Trans. *2LP1790+*
4.366n depredations of those insects] *insert after*: The seeds of Calendula, Marygold, bend up like a hairy caterpillar, with their prickles bristling outwards, and may thus deter some birds or insects from preying upon them. *2LP1790+*
4.366n See note on Rubia] See note on Iris in additional notes; and Cypripedia in Vol. I *2LP1790+*
4.412 bright] fair *emendation? 3LP1791* chill *4LP1794+*

SUMMARY OF BACK MATTER

1LP1789	*2LP1790+*
ADDITIONAL NOTES	**ADDITIONAL NOTES**
Lead the lone pilgrim. [see 1.143n]	*Additional note to Curcuma* [2]. [see Insertions below]
Pleased on the boiling wave. [see 1.353n]	*Addition to the note on Iris*. [see Insertions below]
So shines the glow-fly. [see 4.51n]	*Additional note on Chondrilla*. [see Insertions below]
Rolls his brineless tide. [see 3.343n]	*Addition to the note on Silene* [see Supplement 1–3]
Description of the Upas [see Add'l Notes 1–26]	*Additional note on Ilex*. [see Insertions below]
	Additional note on Ulva. [see Insertions below]
CATALOGUE OF THE POETIC EXHIBITION [not transcribed]	*Additional note on Gossypium* [see 2.93] [*om. 3LP1791+*]
	Addition to Orchis. [see 3.258] [*om. 3LP1791+*]
CONTENTS OF THE NOTES [not transcribed]	*Addition to the note on Tropæolum* [see Supp. 4–16]
	Addition to Avena. [see 4.102] [*om. 3LP1791+*]
INDEX OF THE NAMES OF THE PLANTS [not transcribed]	*Additional note to Bellis*.[see 4.121] [*om. 3LP1791+*]
	Description of the Upas [see Add'l Notes 1–26]
SUPPLEMENT	Account of the Boa Upas [see Add'l Notes 26] [*5LP1799*]

1LP1789	*2LP1790+*
ADDITIONAL NOTES	**ADDITIONAL NOTES**
Addition to the note on Curcuma. [see 1.65n]	**FAIRY-SCENE FROM MR. MUNDY** [see Insertions below] [*3LP1791+*]
Addition to the note on Anthoxanthum. [see 1.86n]	
Addition to the note on Silene [see Supplement 1–3]	**CATALOGUE OF THE POETIC EXHIBITION** [not transcribed]
Addition to the note on Tulipa. [see 1.177n]	
Addition to the note on The Swallow. [see 1.288n]	**CONTENTS OF THE NOTES** [not transcribed]
Addition to the note on Dictamnus. [see 3.184n]	
Addition to the note on Cassia. [see 3.345n]	**INDEX OF THE NAMES OF THE PLANTS** [not transcribed]
Addition to the note on Tropæolum. [see Supp. 4–16]	

ADDITIONAL NOTES

16 ——[*London Mag.*]] *om. 2LP1790+*

26 island.—*London Magazine.*] *insert after*:

Another Account of the Boa Upas, *or* Poison-Tree *of* Macasser, *from an inaugural Dissertation published by* Christ. Aejmelæus, *and approved by* Professor Thunberg, *at* Upsal.

Doctor Aejmelæus first speaks of poisons in general, enumerating many virulent ones from the mineral and animal, as well as from the vegetable kingdoms of Nature. Of the first he mentions arsenical, mercurial, and antimonial preparations; amongst the second he mentions the poisons of several serpents, fishes, and insects; and amongst the last the Curara on the bank of the Oronoko, and the Woorara on the banks of the Amazones, and many others. But he thinks the strongest is that of a tree hitherto undescribed, known by the name of Boa Upas, which grows in many of the warmer parts of India, principally in the islands of Java, Sumatra, Borneo, Bali, Macasser, and Celebes.

Rumphius testifies concerning this Indian poison, that it was more terrible to the Dutch than any warlike instrument; it is by him styled Arbor toxicara, and mentions two species of it, which he terms male and female; and describes the tree as having a thick trunk, with spreading branches, covered with a rough dark bark. The wood, he adds, is very solid, of a pale yellow, and variegated with black spots, but the fructification is yet unknown.

Professor Thunberg supposes the Boa Upas to be a Cestrum, or a tree of the same natural family; and describes a Cestrum of the Cape of Good Hope, the juice of which the Hottentots mix with the venom of a certain serpent, which is said to increase the deleterious quality of them both.

The Boa Upas tree is easily recognised at a distance, being always solitary, the soil around it being barren, and as it were burnt up; the dried juice is dark brown, liquifying by heat, like other refins. it is collected with the greatest caution, the person having his head, hands, and feet carefully covered with linen, that his whole body may be protected from the vapour as well as from the droppings of the tree. No one can approach so near as to gather the juice, hence they supply bamboos, pointed like a spear, which they thrust obliquely, with great force, into the trunk; the juice oozing out gradually fills the upper joint; and the nearer the root wound is made, the more virulent the poison is supposed to be. Sometimes upwards of twenty reeds are left fixed in the tree for three or four days, that the juice may collect and harden in the cavities; the upper joint of the reed is then cut off from the remaining part, the concreted juice is formed into globules or sticks, and is kept in hollow reeds, carefully closed, and wrapped in tenfold linen. It is every week taken out to prevent its becoming mouldy, which spoils it. The deleterious quality appears to be volatile, since it loses much of its power in the time of one year; and in a few years becomes totally effete.

The vapour of the tree produces numbness and spasms of the limbs, and if any one stands under it bare-headed, he loses his hair; and if a drop falls on him, violent inflammation ensues. Birds which sit on the branches a short time, drop down dead, and can even with difficulty fly over it; and not only no vegetables grow under it, but the ground is barren a stone cast around it.

A person wounded by a dart poisoned with this juice feels immediately a sense of heat over his whole body, with great vertigo, to which death soon succeeds. A person wounded with the Java poison was affected with tremor of the limbs, and starting of the tendons in five minutes, and died in less than sixteen minutes, with marks of great anxiety; the corpse, in a few hours, was covered with petechial spots, the face became tumid and lead-coloured, and the white part of the eye became yellow.

The natives try the strength of their poison by a singular test; some of the expressed juice of the root of Amomum Zerumbet is mixed with a little water, and a bit of the poisonous gum or resin is dropped into it; and effervescence instantly takes place, by the violence of which they judge of the strength of the poison.—What air can be extricated during this effervescence?—This experiment is said to be dangerous to the operator.

As the juice is capable of being dissolved in arrack, and is thence supposed to be principally of a resinous nature, the Professor does not credit that fountains have been poisoned with it.

This poison has been employed as a punishment for capital crimes in Macasser and other islands; in those cases some experiments have been made, and when a finger only had been wounded with a dart, the immediate amputation of it did not save the criminal from death.

The poison from what has been termed the femal tree, is less deleterious than the other, and has been used chiefly in hunting; the carcases of animals thus destroyed are eaten with impunity. The poison-juice is said to be used eternally as a remedy against other poisons, in the form of a plaster; also to be used internally for the same purpose; and is believed to alleviate the pain, and extract the poison of venomous insects sooner than any other application.

The author concludes that these accounts have been exaggerated by Mahomedan priests, who have persuaded their followers that the Prophet Mahomet planted this noxious tree as a punishment for the sins of mankind.

An abstract of this Dissertation of C. Aejemalæus is given in Dr. Duncan's Medical Commentaries for the Year 1790, Decad. 2d. Vol. V. *4LP1794+*

SUPPLEMENT

3 thin-body'd] thin-bodied *2LP1790+*

11 *ff.* Lilly] Lily *2LP1790+*

15 *antheræ*] *anthers 2LP1790+*

15 upon the *petals*] *insert after*: Obser. Physique par M. Rozier, Vol. XXXIII. p. 111. *2LP1790+*

16 [*Gen. Evening Post.*]] *om. 2LP1790+*

INSERTIONS TO ADDITIONAL NOTES IN 2LP1790+

Additional note to Curcuma [2]. These anther-less filaments seem to be an endeavour of the plant to produce more stamens, as would appear from some experiments of M. Reynier, instituted for another purpose: he cut away the stamens of many flowers, with design to prevent their fecundity, and in many instances the flower threw out new filaments from the wounded part of different lengths; but did not produce new anthers. The experiments were made on the geum [guem *3LP1791+*] rivale, different kinds of mallows, and the æchinops ritro. Critical Review for March, 1788.

Addition to the note on Iris. In the Persian Iris the end of the lower petal is purple, with white edges and orange streaks, creeping, as it were, into the mouth of the flower like an insect; by which deception in its native climate it probably prevents a similar insect from plundering it of its honey: the edges of the lower petal lap over those of the upper one, which prevents it from opening too wide on fine days, and facilitates its return at night; whence the rain is excluded, and the air admitted. See Polymorpha, Rubia, and Cypripedia in Vol. I.

Additional note on Chondrilla. In the natural state of the expanded flower of the barberry, the stamens lie on the petals; under the concave summits of which the anthers shelter themselves, and in this situation remain perfectly rigid; but on touching the inside of the filament near its base with a fine bristle, or blunt needle, the stamen instantly bends upwards, and the anther, embracing the stigma, sheds its dust. Observations on the Irritation of Vegetables, by T. E. Smith, M. D.

Additional note on Ilex. The efficient cause which renders the hollies prickly in Needwood Forest only as high as the animals can reach them, may arise from the lower branches being constantly cropped by them, and thus shoot forth more luxuriant foliage: it is probable the shears in garden-hollies may produce the same effect, which is equally curious, as prickles are not thus produced on other plants.

Additional note on Ulva. M. Hubert made some observations on the air contained in the cavities of the bambou. The stems of these canes were from 40 to 50 feet in height, and 4 or 5 inches in diameter, and might contain about 30 pints of elastic air. He cut a bambou, and introduced a lighted candle into the cavity, which was extinguished immediately on its entrance. He tried this about 60 times in a cavity of the bambou, containing about two pints. He introduced mice at different times into these cavities, which seemed to be somewhat affected, but soon recovered their agility. The stem of the bambou is not hollow till it rises more than one foot from the earth; the divisions between the cavities are convex downwards. Observ. sur la Physique par M. Rozier, l. 33. p. 130.

INSERTION TO 3LP1791+

Fairy-scene from Mr. Mundy's Needwood Forest. Referred to in Canto IV. l. 35.

HERE, seen of old, the *elfin* race
With sprightly vigils mark'd the place;
Their gay processions charm'd the sight,
Gilding the lucid noon of night;
Or, when obscure the midnight hour,
With glow-worm lantherns hung the bower,
—Hark!—the soft lute!—along the green
Moves with majestic step the QUEEN!
Attendant Fays around her throng,
And trace the dance or raise the song;
Or touch the shrill reed, as they trip,
With finger light and ruby lip.

High, on her brow sublime, is borne
One scarlet wood-bine's tremulous horn;
A gaudy Bee-bird's[1] triple plume
Sheds on her neck its waving gloom;
With silvery gossamer entwin'd
Stream the luxuriant locks behind.
Thin folds of tangled network break
In airy waves adown her neck;—
Warp'd in his loom, the spider spread
The far-diverging rays of thread,

1 The *humming-bird*.

Then round and round with shuttle fine
Inwrought the undulating line;—
Scarce hides the woof her bosom's snow,
One pearly nipple peeps below.
One rose-leaf forms her crimson vest,
The loose edge crosses o'er her breast;
And one translucent fold, that fell
From the tall lily's ample bell,
Forms with sweet grace her snow-white train,
Flows, as she steps, and sweeps the plain.
Silence and Night inchanted gaze,
And Hesper hides his vanquish'd rays!—

Now the wak'd reed-finch swells his throat,
And night-larks trill their mingled note:
Yet hush'd in moss with writhed neck
The blackbird hides his golden beak;
Charm'd from his dream of love, he wakes,
Opes his gay eye, his plumage shakes,
And, stretching wide each ebon wing,
First in low whispers tries to sing;
Then sounds his clarion loud, and thrills
The moon-bright lawns, and shadowy hills.
Silent the choral Fays attend,
And then their silver voices blend,
Each shining thread of sound prolong,
And weave the magic woof of song.
Pleas'd Philomela takes her stand
On high, and leads the Fairy band,
Pours sweet at intervals her strain,
And guides with beating wings the train.
Whilst interrupted Zephyrs bear
Hoarse murmurs from the distant wear;
And at each pause is heard the swell
Of Echo's soft symphonious shell.

Original version in *Needwood Forest* (Lichfield, 1776) [*1776*]

25–26 Scarce hides the ... nipple peeps below] *om. 1776*

29 that] which *1776*

31 snow-white] snowy *1776*

35 waked reed-finch swells his throat] wak'd reed-birds swell their throats *1776*

36 note] notes *1776*

56 symphonious] symphonius *1776*

BIBLIOGRAPHY

I. Works by Erasmus Darwin

The London publisher is Joseph Johnson unless otherwise noted. For select bibliographical descriptions of *The Loves of the Plants*, see Section III below.

Beauties of the Botanic Garden. Ed. Thomas Cadell. London, 1805.

The Botanic Garden, a Poem. In Two Parts.

Part I: Containing The Economy of Vegetation.

1st ed. (***1EV1792***) London, 1792 ["1791"]; rpt. Menston: Scolar P, 1973 (with *1LP1789*); rpt. New York: Garland, 1978 (with *2LP1790*)

2nd ed. (***2EV1792***) London, 1792 ["1791"]

3rd ed. (***3EV1795***) London, 1795

4th ed. (***4EV1799***) London, 1799 (with *5LP1799*)

Part II: The Loves of the Plants.

1st ed. (1LP1789) Lichfield, 1789; rpt. Menston: Scolar P, 1973 (with *1EV1792*); rpt. Oxford: Woodstock Books, 1991

2nd ed. (***2LP1790***) London, 1790; rpt. New York: Garland, 1978 (with *1EV1792*)

3rd ed. (***3LP1791***) London, 1791

4th ed. (***4LP1794***) London, 1794

5th ed. (***5LP1799***) London, 1799 (with *4EV1799*)

The Botanic Garden, a Poem. In Two Parts.

Part I: Containing The Economy of Vegetation (Dublin, 1793).

Part II: The Loves of the Plants

1st ["3rd"] ed. Dublin, 1790

2nd ["4th"] ed. Dublin ("Dudlin"), 1796

The Botanic Garden, a Poem. In Two Parts. Part I: Containing The Economy of Vegetation. Part II: The Loves of the Plants. New York, 1798; 2nd ed. 1807.

The Botanic Garden, a Poem. In Two Parts. Part I: Containing The Economy of Vegetation. Part II: The Loves of the Plants. The Temple of Nature, or The Origin of Society. London: Jones, 1824.

The Collected Letters of Erasmus Darwin. Ed. Desmond King-Hele. New York: Cambridge University Press, 2007.

The Collected Writings of Erasmus Darwin. Ed. Martin Priestman. 9 vols. Bristol: Thoemmes Continuum, 2004.

The Essential Writings of Erasmus Darwin. Ed. Desmond King-Hele. London: MacGibbon & Kee, 1968.

The Families of Plants. Trans. Genera plantarum by Carl von Linné. 2 vols. London, 1787.

"The Folly of Atheism." Rpt. *The Monthly Magazine* 13.87 (June 1802): 452–53.

Phytologia; or, The Philosophy of Agriculture and Gardening. London, 1800.

A Plan for the Conduct of Female Education, in Boarding Schools. 1794. London, 1797.

The Poetical Works of Erasmus Darwin. 3 vols. London, 1806.

A System of Vegetables. Trans. *Systema vegetabilium* by Carl von Linné. 2 vols. London, 1783.

The Temple of Nature; or, The Origin of Society. London, 1803. Rpt. ed. Martin Priestman. Romantic Circles, 2006. Web. <www.rc.umd.edu/editions/darwin_temple/>.

Zoonomia; or, The Laws of Organic Life. Part I, London, 1794 (2nd ed. London, 1796); Parts II and III, London, 1796. Rpt. New York: AMS, 1974.

II. Works by Other Authors

Works included here are principally those cited in the Editors' Introduction or throughout the edition. For other works cited, see Explanatory Notes, Textual Notes and Descriptions of the Illustrations. For further information, see www.thebotanicgarden.net.

Addison, Joseph. "The Pleasures of the Imagination." Paper 1. *The Spectator* 411 (21 June 1712): 823–24.

Bewell, Alan. "'Jacobin Plants:' Botany as Social Theory." *The Wordsworth Circle* 20.3 (1989): 132–139.

——."Erasmus Darwin's Cosmopolitan Nature." *ELH* 76.1 (2009): 19–48.

"Biographical Memoirs of the Late Dr. Darwin." *The Monthly Magazine* 13.87 (1802): 457–63.

Blake, William. *The Complete Poetry and Prose of William Blake*. Rev. ed. David V. Erdman. Commentary by Harold Bloom. New York: Anchor, 1988.

Boime, Albert. *Art in an Age of Revolution, 1750–1800*. Chicago: University of Chicago Press, 1987.

Browne, Janet. "Botany for Gentlemen: Erasmus Darwin and 'The Loves of the Plants.'" *Isis* 80.4 (1989): 592–621.

Brylowe, Thora. "Antiquity by Design: Re-Mediating the Portland Vase." *Romantic Antiquarianism*. Ed. Noah Heringman and Crystal Lake. Romantic Circles, 2014. Web. <www.rc.umd.edu/praxis/antiquarianism/praxis.antiquarianism.2014.brylowe.html>.

Budge, Gavin. "Erasmus Darwin and the Poetics of William Wordsworth: 'Excitement without the Application of Gross and Violent Stimulants.'" *[British] Journal for Eighteenth-Century Studies* 30.2 (2007): 279–308.

Calè, Luisa, and Adriana Craciun. "The Disorder of Things." *Eighteenth-Century Studies* 45.1 (2011): 1-13.

Canning, George, and John Frere. *Poetry of the Anti-Jacobin. Revolution and Romanticism, 1789–1834*. New York: Woodstock Books, 1991.

Chambers, Ephraim. *Cyclopaedia, or, an Universal Dictionary of Arts and Sciences*. 2 vols. London, 1728.

Coffey, Donna. "Protecting the Botanic Garden: Seward, Darwin and Coalbrookdale." *Women's Studies* 31.2 (2002): 141–64.

Coleridge, Samuel Taylor. *The Collected Works of Samuel Taylor Coleridge*. Ed. James Engell, W. Jackson Bate et al. 16 vols. Princeton: Princeton UP, 1969–2001.

——. *Collected Letters of Samuel Taylor Coleridge*. Ed. Earl Leslie Griggs. 6 vols. Oxford: Clarendon P, 2002.

Cowper, William. *The Letters and Prose Writings of William Cowper*. 5 vols. Ed. James King and Charles Ryskamp. New York: Oxford University Press, 1979–86.

D'Alembert, Jean Le Rond. *Preliminary Discourse to the Encyclopedia of Diderot*. Trans. Richard N. Schwab. Chicago: University of Chicago Press, 1995.

Darwin, Charles. *The Life of Erasmus Darwin*. Ed. Desmond King-Hele. New York: Cambridge University Press, 2003.

——. *The Descent of Man, and Selection in Relation to Sex*. New York: Penguin, 2004.

——. *On the Origin of Species*. New York: Penguin, 2009.

de Almeida, Hermione. "Romanticism and the Triumph of Life Science: Prospects for Study." *Studies in Romanticism* 43.1 (2004): 119–34.

Dushane, Allison. " 'Mere Matter: Causality, Subjectivity and Aesthetic Form in Erasmus Darwin." Romanticism and Victorianism on the Net 56 (2009). Web < www.erudit.org/revue/ravon/2009/v/n56/1001100ar.html>.

Engelstein, Stefani. *Anxious Anatomy: The Conception of the Human Form in Literary and Naturalist Discourse*. Albany: State University of New York Press, 2008.

Fara, Patricia. *Erasmus Darwin: Sex, Science, and Serendipity*. Oxford: Oxford University Press, 2012.

Ford, Jennifer. *Coleridge on Dreaming: Romanticism, Dreams and the Medical Imagination*. New York: Cambridge University Press, 1998.

Foucault, Michel. *The Order of Things: An Archaeology of the Human Sciences*. 1970. Trans. of *Le mots et les choses*. 1966. New York: Vintage, 1994.

Fulford, Tim. "Coleridge, Darwin, Linnaeus: The Sexual Politics of Botany." *The Wordsworth Circle* 28.3 (1997): 124–30.

Fulford, Tim, Debbie Lee and Peter Kitson. *Literature, Science and Exploration in the Romantic Era: Bodies of Knowledge*. New York: Cambridge University Press, 2004.

George, Sam. *Botany, Sexuality and Women's Writing 1760–1830: From Modest Shoot to Forward Plant*. New York: Manchester University Press, 2007.

Gigante, Denise. *Life: Organic Form and Romanticism*. New Haven: Yale University Press, 2009.

Golinski, Jan. *Science as Public Culture. Chemistry and Enlightenment in Britain, 1760–1820*. New York: Cambridge University Press, 1992.

Green, Matthew. "Blake, Darwin and the Promiscuity of Knowing: Rethinking Blake's Relationship to the Midlands Enlightenment." *[British] Journal for Eighteenth-Century Studies* 30.2 (2007): 193–208.

Heringman, Noah. *Romantic Rocks, Aesthetic Geology*. Ithaca, NY: Cornell University Press, 2004.

Hunt, Leigh. *The Feast of the Poets*. London, 1814.

Hunter, Allan. "Evolution, Revolution and Frankenstein's Creature." *Frankenstein's Science: Experimentation and Discovery in Romantic Culture, 1780–1830*. Ed. Christa Knellwolf. Burlington, VT: Ashgate, 2008. 133–49.

Hutton, James. Letter to George Clerk-Maxwell, August 1774. "The Correspondence between James Hutton (1726-1797) and James Watt (1736-1819) with Two Letters from Hutton to George Clerk-Maxwell (1715-1784): Part I." By Jean Jones, Hugh Torrens and Eric Robinson. *Annals of Science* 51.6 (1994): 637–53.

Jackson, Noel. *Science and Sensation in Romantic Poetry*. New York: Cambridge University Press, 2008.

Kelley, Theresa. "Romantic Nature Bites Back: Adorno and Romantic Natural History." *European Romantic Review* 15.2 (2004): 193–203.

——. *Clandestine Marriage: Botany and Romantic Culture*. Baltimore: Johns Hopkins University Press, 2012.

King, Amy. *Bloom: The Botanical Vernacular in the English Novel*. New York: Oxford University Press, 2003.

King-Hele, Desmond. *Erasmus Darwin*. New York: Scribners, 1964.

——. *Doctor of Revolution: The Life and Genius of Erasmus Darwin*. London: Faber & Faber, 1977.

——. *Erasmus Darwin and the Romantic Poets*. New York: St. Martin's Press, 1986.

——. *Erasmus Darwin: A Life of Unequalled Achievement*. London: DLM, 1999.

Klancher, Jon. *Transfiguring the Arts and Sciences: Knowledge and Cultural Institutions in the Romantic Age*. New York: Cambridge University Press, 2013.

Komisaruk, Adam. "Mystifying What Matters: Erotic Antiquarianism in Erasmus Darwin's Portland Vase." *Eighteenth-Century Life* 38.3 (2014): 1–29.

Krause, Ernst. *Erasmus Darwin*. Trans. W.S. Dallas. Preliminary Notice by Charles Darwin. London, 1879.

Liu, Lydia. "Robinson Crusoe's Earthenware Pot: Science, Aesthetics and the Metaphysics of True Porcelain." *Romantic Science: The Literary Forms of Natural History*. Ed. Noah Heringman. Albany: State U of New York P, 2003. 139–71.

"The Loves of the Triangles. A Mathematical and Philosophical Poem. Inscribed to Dr. Darwin." *The Anti-Jacobin; or, Weekly Examiner* 23 (16 April 1798): 162–74; 24 (23 April 1798): 200–05. Rpt. *Poetry of the Anti-Jacobin*. London, 1799. 108–29.

Mahood, M.M. *The Poet as Botanist*. New York: Cambridge University Press, 2008.

Mathias, Thomas James. *The Pursuits of Literature: A Satirical Poem in Four Dialogues*. 1794–97. 8th ed. London, 1798.

Matlak, Richard. *The Poetry of Relationship: The Wordsworths and Coleridge, 1797–1800*. New York: St. Martin's Press, 1997.

McGann, Jerome. *The Poetics of Sensibility: A Revolution in Literary Style*. New York: Oxford University Press, 1996.

McNeil, Maureen. *Under the Banner of Science: Erasmus Darwin and His Age*. Wolfeboro, N.H.: Manchester University Press, 1987.

Mellor, Anne. *Mary Shelley: Her Life, Her Fiction, Her Monsters*. New York: Routledge, 1989.

Moore, Lisa. Sister Arts: *The Erotics of Lesbian Landscapes*. Minneapolis: Minnesota University Press, 2011.

Morton, Timothy. *The Poetics of Spice: Romantic Consumerism and the Exotic*. New York: Cambridge University Press, 2000.

Mundy, Francis Noel Clarke. *Needwood Forest*. Lichfield, 1776.

Nichols, Ashton, ed. *Romantic Natural Histories: Selected Texts with Introduction*. New York: Houghton Mifflin, 2004.

——, ed. *Romantic Natural History*. Dickinson College, 8 August 2007. Web. 18 May 2011. <http://users.dickinson.edu/~nicholsa/Romnat/>.

Ogden, Daryl. *The Language of the Eyes: Science, Sexuality and Female Vision in English Literature and Culture, 1690–1927*. Albany: State University of New York Press, 2005.

"P.P." [William Cowper]. Rev. of *The Botanic Garden, Part I: The Economy of Vegetation*, by Erasmus Darwin. *Analytical Review* 15.3 (March 1793): 287–93.

Packham, Catherine. "The Science and Poetry of Animation: Personification, Analogy and Erasmus Darwin's *Loves of the Plants*." *Romanticism* 10.2 (2004): 191–208.

Page, Michael. *The Literary Imagination from Erasmus Darwin to H.G. Wells: Science, Evolution and Ecology*. Burlington: Ashgate, 2012.

Plumptre, James. *The Lakers: A Comic Opera, in Three Acts*. London, 1798.

Polwhele, Richard. "The Unsex'd Females: A Poem." 1798. New York, 1800.

Porter, Dahlia. "Scientific Analogy and Literary Taxonomy in Darwin's *Loves of the Plants*." *European Romantic Review* 18.2 (2007): 213–21.

The Portland Vase. Spec. issue of *Journal of Glass Studies* 32 (1990).

Priestman, Martin. *Romantic Atheism: Poetry and Freethought, 1780–1830*. New York: Cambridge University Press, 1999.

——. "Lucretius in Romantic and Victorian Britain." *The Cambridge Companion to Lucretius*. New York: Cambridge University Press, 2010. 289–305.

——. *The Poetry of Erasmus Darwin: Enlightened Spaces, Romantic Times*. Burlington, VT: Ashgate, 2013.

Reed, Edward. *From Soul to Mind: The Emergence of Psychology from Erasmus Darwin to William James*. New Haven: Yale University Press, 1997.

Reno, Seth T. "Rethinking the Romantic's Love of Nature." *Wordsworth and the Green Romantics: Affect and Ecology in the Nineteenth Century*. Ed. Lisa Ottum and Seth T. Reno. Burlington, VT: U of New Hampshire P, 2016. 28–58.

Rev. of *The Botanic Garden*, by Erasmus Darwin. *The English Review of Literature, Science, Discoveries, Inventions and Practical Controversies and Contests* 20 (September 1792): 161–71.

Rev. of *The Botanic Garden, Part I: The Economy of Vegetation*, by Erasmus Darwin. *Critical Review* 6 (October 1792): 162–71.

Rev. of *The Botanic Garden, Part II: The Loves of the Plants*, by Erasmus Darwin. *English Review of Literature, Science, Discoveries, Inventions and Practical Controversies and Contests* 14 (August 1789): 128–32.

Rev. of *Zoonomia*, by Erasmus Darwin. *The British Critic, and Quarterly Theological Review* 5 (February 1795): 113–22.

Richardson, Alan. *British Romanticism and the Science of the Mind*. New York: Cambridge University Press, 2001.

Rousseau, George. "'Brainomania': Brain, Mind and Soul in the Long Eighteenth Century." *[British] Journal for Eighteenth-Century Studies* 30.2 (2007): 161–91.

Rousseau, G.S. and Roy Porter, eds. *The Ferment of Knowledge: Studies in the Historiography of Eighteenth-Century Science*. New York: Cambridge University Press, 1980.

Ruston, Sharon. *Shelley and Vitality*. New York: Palgrave, 2005.

Schiebinger, Londa. *Nature's Body: Gender in the Making of Modern Science*. New Brunswick, NJ: Rutgers University Press, 2004.

Seligo, Carlos. "The Monsters of Botany and Mary Shelley's *Frankenstein*." *Science Fiction: Critical Frontiers*. Ed. Karen Sayer and John Moore. New York: St. Martin's Press, 2000. 69–84.

Setzer, Sharon. "'Pond'rous Engines' in 'Outraged Groves': The Environmental Argument of Anna Seward's 'Colebrook Dale.'" *European Romantic Review* 18.1 (2007): 69–82.

Seward, Anna. *Memoirs of the Life of Dr. Darwin*. London, 1804.

——. *The Poetical Works of Anna Seward; with Extracts from Her Literary Correspondence*. 3 vols. Ed. Walter Scott. London, 1810.

Sha, Richard. *Perverse Romanticism: Aesthetics and Sexuality in Britain, 1750–1832*. Baltimore: Johns Hopkins University Press, 2009.

Shelley, Mary. *Frankenstein; or, The Modern Prometheus*. 1818. Ed. Maurice Hindle. New York: Penguin, 2003.

Shteir, Ann. *Cultivating Women, Cultivating Science: Flora's Daughters and Botany in England, 1760–1860*. Baltimore: Johns Hopkins University Press, 1996.

Smith, C.U.M., and Robert Arnott, eds. *The Genius of Erasmus Darwin*. Burlington, VT: Ashgate, 2005.

Sterne, Laurence. *The Life and Opinions of Tristram Shandy, Gentleman*. 9 vols. London, 1759–67.

Stevens, William Bagshaw. *The Journal of the Rev. William Bagshaw Stevens*. Ed. Georgina Galbraith. Oxford: Clarendon Press, 1965.

Stukley, William. "An Account of the Impression of the Almost Entire Sceleton of a Large Animal in a Very Hard Stone, Lately Presented the Royal Society, from Nottinghamshire." *Philosophical Transactions* 30.360 (1717–19): 963–68.

Teute, Frederika. "The Loves of the Plants; or, the Cross-Fertilization of Science and Desire at the End of the Eighteenth Century." *British Radical Culture of the 1790s*. Ed. Robert Maniquis. San Marino, CA: Huntington Library, 2002. 63–89.

Trott, Nicola. "Wordsworth's Loves of the Plants." *1800: The New Lyrical Ballads*. Ed. Nicola Trott and Seamus Perry. New York: Palgrave, 2001. 141–68.

Uglow, Jenny. *The Lunar Men: Five Friends Whose Curiosity Changed the World*. New York: Farrar, Straus and Giroux, 2002.

Vickers, Neil. *Coleridge and the Doctors, 1795–1806*. New York: Oxford University Press, 2004.

Wordsworth, William. *Complete Poetical Works*. Ed. Thomas Hutchinson. Rev. ed. Ernest De Selincourt. New York: Oxford University Press, 1936.

Yeo, Richard. "Classifying the Sciences." *Eighteenth-Century Science*. Vol. 4 of *The Cambridge History of Science*. New York: Cambridge University Press, 2003.

III. Select Bibliographical Descriptions: *The Loves of the Plants*

The following describe the British Library copies (with English Short Title Catalogue numbers) of the English lifetime editions of *The Loves of the Plants*, digitized in *Eighteenth-Century Collections Online* and checked against physical copies in the Folger Shakespeare Library (Washington, DC), the Harry Ransom Center (University of Texas, Austin) and the Library of Congress.

1LP1789 **(ESTC T082159)**

THE | BOTANIC GARDEN, | [INSERTED?] PART II. | CONTAINING | The LOVES of the PLANTS, | A | POEM. | WITH | PHILOSOPHICAL NOTES. | [rule] | VOLUME THE SECOND. | VIVUNT IN VENEREM FRONDES; NEMUS OMNE PER ALTUM | FELIX ARBOR AMAT; NUTANT AD MUTUA PALMÆ | FÆDERA, POPULEO SUSPIRAT POPULUS ICTU, | ET PLATANI PLATANIS, ALNOQUE ASSIBILAT ALNUS. | CLAUD. EPITH. | [DOUBLE RULE] | *LICHFIELD:* PRINTED BY *J. JACKSON.* | Sold by J. Johnson, St. Paul's Church Yard, London: | M, DCC, LXXXIX. | [Entered at Stationer's Hall.]

4°: π^4 A^4 B-D^4 E^1 E-Z^4 2A-$2B^2$ χ^1 [$2 signed]; 102 leaves, pp. [4] v–viii [10] 1–24 24–25 25–184 [2] [=204]; 8 plates

Contents: *i* short title *iii* title v PROEM. [*ix*] ERRATA. [*xi*] ADVERTISEMENT. [*xiii*] PREFACE. *1* text 166 ADDITIONAL NOTES. 174 CATALOGUE | OF THE | POETIC EXHIBITION. 176 CONTENTS OF THE NOTES. 179 INDEX OF THE NAMES OF THE PLANTS. 181 SUPPLEMENT. [*185*] Directions to the binder; ERRATA.

2LP1790 **(ESTC T082157)**

THE | BOTANIC GARDEN. | PART II. | CONTAINING | THE LOVES OF THE PLANTS. | A POEM. | with | PHILOSOPHICAL NOTES. | [single rule] | VOLUME THE SECOND. | [single rule] | [Claudian epigraph] | [single rule] | THE SECOND EDITION. | [single rule] | LONDON: | PRINTED BY J. NICHOLS, | FOR J. JOHNSON, ST. PAUL'S CHURCH YARD. | M, DCC, XC. | [Entered at Stationer's Hall.]

4°: A^2 B^4-$2E^4$ [$2 signed]; 110 leaves, pp. [6] iii–xii 1–202 [2] [=215]; 10 plates

Contents: short title title *i* advertisement *iii* preface ix proem *1* text 185 additional notes 195 catalogue of the poetic exhibition 197 contents of the notes 201 index of the names of the plants *203* directions to the binder; errata *204* Lately Published by J. Johnson, in St. Paul's Church Yard.

3LP1791 **(ESTC T082156)**

THE | BOTANIC GARDEN. | PART II. | CONTAINING | THE LOVES OF THE PLANTS. | *A POEM.* | WITH | Philoſophical Notes. | [double rule] | [Claudian epigraph] | [double rule] | [double rule] | THE THIRD EDITION. | [double rule] | LONDON, | PRINTED FOR J. JOHNSON, ST. PAUL's CHURCH-YARD. | [dots] | MDCCXCI. | Entered at Stationers Hall.

4°: π^2 A-$2C^4$ [$2 (-B1, K2, L2, N2, U2) signed]; 106 leaves, pp. [4] i–x [vi misnumbered iv] 1–198 [=212]; 10 plates

Contents: short title title *i* preface *vii* proem 1 text *181* additional notes 189 *Fairy-Scene from Mr. Mundy's Needwood Forest. Referred to in Canto IV. l. 35.* 191 catalogue of the poetic exhibition 193 contents of the notes 195 index of the names of the plants 197 errata; emendations; directions to the binder

4LP1794 **(ESTC T082155)**

THE | BOTANIC GARDEN. | PART II. | CONTAINING | THE LOVES OF THE PLANTS. | *A POEM.* | WITH | Philoſophical Notes. | [double rule] | [Claudian epigraph] | [double rule] | [double rule] | THE FOURTH EDITION. | [double rule] | LONDON, | PRINTED FOR J. JOHNSON, ST. PAUL's CHURCH-YARD. | [dots] | MDCCXCIV. | *ENTERED AT STATIONERS' HALL.*

4°: π^2 A-$2C^4$ χ^1 [$2 (-R2, 2C2) signed]; 107 leaves, pp. [4] i–viii 1–200 [2] [=214]; 10 plates

Contents: short title volume title *i* preface *vii* proem *1* text *183* additional notes 193 Mundy fairy-scene 195 catalogue of the poetic exhibition 197 contents of the notes 199 index of the names of the plants *201* directions to the binder

***5LP1799* (ESTC T146846)**

WORK TITLE:

THE BOTANIC GARDEN, | A POEM. | IN TWO PARTS. | [double rule] | PART I. | CONTAINING | THE ECONOMY OF VEGETATION. | [double rule] | PART II. | THE LOVES OF THE PLANTS. | [double rule] | WITH | PHILOSOPHICAL NOTES. | [double rule] | THE FOURTH EDITION. | [double rule] | LONDON: | PRINTED FOR J. JOHNSON, ST. PAUL'S CHURCH-YARD. | [single rule] | 1799.

VOLUME TITLE:

THE | BOTANIC GARDEN. | PART II. | CONTAINING | THE LOVES OF THE PLANTS, | A POEM. | WITH | PHILOSOPHICAL NOTES. | [swelled rule] | [Claudian epigraph]

8°: π^2 [-π2] A^6 B-S^8 T^4 U^2 [$4 signed]; 149 leaves, pp. [4] iii-xii 1–282 [2] [=298]; 12 plates
Contents: volume title *i* work title *iii* preface *1* text *239* additional notes 266 Mundy fairy-scene 269 catalogue of the poetic exhibition *273* contents of the notes 281 index of the names of the plants [*283*] AN ADDITION, | *To be inserted near the end of the Additional Note* XXXIII ... *of the first volume, immediately before the last sentence*; directions to the binder; erratum

IV. Additional Works Consulted

Andersen, Geoff. *Telescope: Its History, Technology and Future*. Princeton: Princeton University Press, 2007.

Anglesey, George Charles Henry Victor Paget, Marquis of. *One-Leg: The Life and Letters of Henry William Paget, First Marquess of Anglesey*. New York: Morrow, 1961.

Baird, Davis. *Thing Knowledge: A Philosophy of Scientific Instruments*. Berkeley: University of California Press, 2004.

Barrell, John. *The Political Theory of Painting from Reynolds to Hazlitt: 'The Body of the Public.'*" New Haven: Yale University Press, 1986.

Bassiri, Nima. "The Brain and the Unconscious Soul in Eighteenth-Century Nervous Physiology: Robert Whytt's Sensorium Commune." *Journal of the History of Ideas* 74.3 (2013): 425–48.

Bellos, Alex. *Here's Looking at Euclid: A Surprising Excursion through the Astonishing World of Math*. New York: Free Press, 2010.

Bevan, John. "Diving Bells through the Centuries." *South Pacific Underwater Medicine Society Journal* 29.1 (1999): 42–50.

"Biographical Sketches of Eminent Inventors: 34." *The Technical Educator* 3 (1871): 30–31.

Blunt, Wilfrid, with William Stearn. *Linnaeus: The Compleat Naturalist*. Princeton: Princeton University Press, 2001.

Bowers, Toni. *The Politics of Motherhood: British Writing and Culture, 1680–1760*. New York: Cambridge University Press, 1996.

Bowler, Peter, and Iwan Morus. *Making Modern Science: A Historical Survey*. Chicago: University of Chicago Press, 2005.

Briggs, J. Morton. "Aurora and Enlightenment: Eighteenth-Century Explanations of the Aurora Borealis." *Isis* 58.4 (1967): 491–503.

Brody, Inger. *Ruined by Design: Shaping Novels and Gardens in the Culture of Sensibility*. New York: Routledge, 2008.

Brown, Laura. *Homeless Dogs and Melancholy Apes: Humans and Other Animals in the Modern Literary Imagination*. Ithaca: Cornell University Press, 2010.

Bulfinch, Thomas. *The Age of Fable*. Boston, 1855.

Bynum, W.F., and Roy Porter, eds. *Companion Encyclopedia of the History of Medicine*. 2 vols. New York: Routledge, 1993.

Cantor, G.N. *Optics after Newton: Theories of Light in Britain and Ireland, 1704–1840*. Dover, NH: Manchester University Press, 1983.

Chang, Hasok. *Inventing Temperature: Measurement and Scientific Progress*. New York: Oxford University Press, 2007.

Close, F.E. *The Void*. New York: Oxford University Press, 2010.

Cohen, I. Bernard. *Benjamin Franklin's Science*. Cambridge, MA: Harvard University Press, 1990.

Coogan, Michael, ed. *The Oxford Encyclopedia of the Books of the Bible*. New York: Oxford University Press, 2011.

Crone, Robert. *A History of Color: The Evolution of Theories of Lights and Color*. Norwell, MA: Kluwer, 1999.

Dacome, Lucia. "'To What Purpose Does It Think?': Dreams, Sick Bodies and Confused Minds in the Age of Reason." *History of Psychiatry* 15.4 (2004): 395–416.

Dray, Philip. *Stealing God's Thunder: Benjamin Franklin's Lightning Rod and the Invention of America*. New York: Random House, 2005.

During, Simon. *Modern Enchantments: The Cultural Power of Secular Magic*. Cambridge, MA: Harvard University Press, 2002.

Eales, Nellie. "The History of the Lymphatic System, with Special Reference to the Hunter–Monro Controversy." *Journal of the History of Medicine and Allied Sciences* 29.3 (1974): 280–94.

Emsley, John. *The 13th Element: The Sordid Tale of Murder, Fire and Phosphorus*. New York: Wiley, 2000.

Fagan, Brian. *Elixir: A History of Water and Humankind*. New York: Bloomsbury, 2011.

Fara, Patricia. *An Entertainment for Angels: Electricity in the Enlightenment*. New York: Columbia University Press, 2002.

——. "Lord Derwentwater's Lights: Prediction and the Aurora Polaris." *Journal for the History of Astronomy* 27.3 (1996): 239–58.

——. *Science: A Four Thousand Year History*. New York: Oxford University Press, 2010.

Feilla, Cecilia. "From 'Sainted Maid' to 'Wife in All Her Grandeur': Translations of Heloise, 1687–1817." *Eighteenth-Century Life* 28.2 (2004): 1–16.

Finger, S., F. Boller, and K.L. Tyler, eds. *History of Neurology*. Vol. 95, 3rd series, of *Handbook of Clinical Neurology*. Amsterdam: Elsevier, 2014.

Finn, Bernard. "Laplace and the Speed of Sound." *Isis* 55.179 (1964): 7–19.

Fitton, R.S. *The Arkwrights: Spinners of Fortune*. New York: St. Martin's Press, 1989.

Francis, Michael, ed. *Iberia and the Americas: Culture, Politics and History*. Santa Barbara, CA: ABC–CLIO, 2006.

Frisinger, H. Howard. *The History of Meteorology to 1800*. New York: Science History Publications, 1977.

Fulford, Tim. *Landscape, Liberty and Authority: Poetry, Criticism and Politics from Thomson to Wordsworth*. 1996. New York: Cambridge University Press, 2006.

Genz, Henning. *Nothingness: The Science of Empty Space*. Trans. Karin Heusch. New York: Basic Books, 2001.

Gigante, Denise. *Taste: A Literary History*. New Haven: Yale University Press, 2005.

Gillispie, Charles. *The Montgolfier Brothers and the Invention of Aviation, 1783–1784*. Princeton: Princeton University Press, 1983.

——. *Science and Polity in France: The End of the Old Regime*. Princeton: Princeton University Press, 2004.

Gilmour, Rodney. *Phosphoric Acid: Purification, Uses, Technology and Economics*. Boca Raton, FL: CRC Press, 2014.

Global Volcanism Program. Department of Mineral Sciences, National Museum of Natural History, Smithsonian Institution, 2013. Web <volcano.si.edu>.

Goldstein, Bernard, and Giora Hon. "Kepler's Move from Orbs to Orbits: Documenting a Revolutionary Scientific Concept." *Perspectives on Science* 13.1 (2005): 74–111.

Golinski, Jan. "Barometers of Change: Meteorological Instruments as Machines of Enlightenment." *The Sciences in Enlightened Europe*. Ed. William Clark, Jan Golinski and Simon Schaffer. Chicago: University of Chicago Press, 1999. 69–93.

——. *British Weather and the Climate of Enlightenment*. Chicago: University of Chicago Press, 2007.

Graves, Robert. *The Greek Myths*. 1955. New York: Penguin, 2012.

Greenblatt, Stephen. *Swerve: How the World Became Modern*. New York: Norton, 2011.

Hamilton, Edith. *Mythology: Timeless Tales of Gods and Heroes*. 1942. New York: Back Bay, 1998.

Hankins, Thomas. "The Ocular Harpsichord of Louis-Bertrand Castel; or, The Instrument That Wasn't." *Osiris* 9.1 (1994): 141–56.

Hannings, Bud. *Chronology of the American Revolution: Military and Political Actions Day by Day*. Jefferson, NC: McFarland, 2008.

Harvey, E. Newton. *History of Luminescence: From the Earliest Times until 1900*. 1957. Mineola, NY: Dover, 2005.

Haskell, Francis, and Nicholas Penny. *Taste and the Antique: The Lure of Classical Sculpture, 1500–1900*. New Haven: Yale University Press, 1981.

Heilbron, J.L. *Electricity in the 17th and 18th Centuries: A Study of Early Modern Physics*. 1979. Mineola, NY: Dover, 1999.

Henry, John. "Gravity and *De gravitatione*: The Development of Newton's Ideas on Action at a Distance." *Studies in the History and Philosophy of Science* 42.1 (2011): 11–27.

Hesse, Mary. *Forces and Fields: The Concept of Action at a Distance in the History of Physics*. 1961. Mineola, NY: Dover, 2005.

Hills, Richard. *Power from Steam: A History of the Stationary Steam Engine*. New York: Cambridge University Press, 1989.

Hoeppe, Götz. *Why the Sky Is Blue: Discovering the Color of Life*. Trans. with John Stewart. Princeton: Princeton University Press, 2007.

Hopkins, Donald. *The Greatest Killer: Smallpox in History*. Chicago: University of Chicago Press, 2002.

Hunt, L.B. "The True Story of Purple of Cassius." *Gold Bulletin* 9.4 (1976): 134–39.

Iliffe, Rob. "Abstract Considerations: Disciplines and the Incoherence of Newton's Natural Philosophy." *Studies in the History and Philosophy of Science* 35.3 (2004): 427–54.

Isler, Hansruedi. "The Development of Neurology and the Neurological Sciences in the 17th Century." Finger et al. 91–106.

Jago, Lucy. *The Northern Lights.* New York: Knopf, 2001.

Jankovic, Vladimir. *Reading the Skies: A Cultural History of English Weather, 1650–1820.* Chicago: University of Chicago Press, 2001.

Jobes, Gertrude. *Dictionary of Mythology, Folklore and Symbols.* 3 vols. New York: Scarecrow Press, 1961–62.

Kavanagh, Thomas. *Enlightened Pleasures: Eighteenth-Century France and the New Epicureanism.* New Haven: Yale University Press, 2010.

Keen, Paul. "The 'Balloonomania': Science and Spectacle in 1780s England." *Eighteenth-Century Studies* 39.4 (2006): 507–35.

Kilfeather, Siobhán. *Dublin: A Cultural History.* New York: Oxford University Press, 2005.

King, Henry. *History of the Telescope.* 1955. Mineola, NY: Dover, 2003.

Kipp, Julie. *Romanticism, Maternity and the Body Politic.* New York: Cambridge University Press, 2003.

Kitson, Peter. *Forging Romantic China: Sino-British Cultural Exchange 1760–1840.* New York: Cambridge University Press, 2013.

Klein, Maury. *The Power Makers: Steam, Electricity and the Men Who Invented Modern America.* New York: Bloomsbury, 2008.

Klein, Ursula, and Wolfgang Lefèvre. *Materials in Eighteenth-Century Science: A Historical Ontology.* Cambridge, MA: MIT Press, 2007.

Kohn, George, ed. *Encyclopedia of Plague and Pestilence.* New York: Facts on File, 2008.

Kowaleski-Wallace, Beth. "Tea, Gender and Domesticity in Eighteenth-Century England." *Studies in Eighteenth-Century Culture* 23 (1994): 131–45.

Krider, E. Philip. "Benjamin Franklin and Lightning Rods." *Physics Today* 59 (2006): 42–48.

Kuhn, Thomas. *The Structure of Scientific Revolutions.* 1962. Chicago: University of Chicago Press, 2012.

Leapman, Michael. *The Ingenious Mr. Fairchild: The Forgotten Father of the Flower Garden.* New York: St. Martin's Press, 2001.

Lemay, J.A. Leo. *The Life of Benjamin Franklin.* 3 vols. Philadelphia: University of Pennsylvania Press, 2006–09.

Levere, Trevor. *Transforming Matter: A History of Chemistry from Alchemy to the Buckyball.* Baltimore: Johns Hopkins University Press, 2001.

Leverington, David. *Babylon to Voyager and Beyond: A History of Planetary Astronomy.* New York: Cambridge University Press, 2003.

Loadman, John. *Tears of the Tree: The Story of Rubber—A Modern Marvel.* New York: Oxford University Press, 2005.

Looby, Christopher. "Phonetics and Politics: Franklin's Alphabet as a Political Design." *Eighteenth-Century Studies* 18.1 (1984): 1–34.

McClellan, James. "The Scientific Press in Transition: Rozier's Journal and the Scientific Societies in the 1770s." *Annals of Science* 36.5 (1979): 425–49.

Maniquis, Robert. "The Puzzling Mimosa: Sensitivity and Plant Symbols in Romanticism." *Studies in Romanticism* 8 (1969): 129–55.

Markham, Clements. *Peruvian Bark: A Popular Account of the Introduction of Chinchona Cultivation into British India*. London, 1880.

Mazzeo, Tilar. *Plagiarism and Literary Property in the Romantic Period*. Philadelphia: University of Pennsylvania Press, 2007.

Miall, David. "The Meaning of Dreams: Coleridge's Ambivalence." *Studies in Romanticism* 21.1 (1982): 57–71.

Moore, Lisa. "Queer Gardens: Mary Delany's Flowers and Friendships." *Eighteenth-Century Studies* 39.1 (2005): 49–70.

Moote, A. Lloyd, and Dorothy Moote. *The Great Plague: The Story of London's Most Deadly Year*. Baltimore: Johns Hopkins University Press, 2004.

Morris, Richard. *The Last Sorcerers: The Path from Alchemy to the Periodic Table*. Washington: Joseph Henry Press, 2003.

Murray, John. *Memoir on the Diamond, Including Its Economical and Political History*. London, 1839.

Musson, A.E., and Eric Robinson. *Science and Technology in the Industrial Revolution*. Toronto: University of Toronto Press, 1969.

Neis, Cordula. "European Conceptions of Writing from the Renaissance to the Eighteenth Century." *History of Linguistics 2008: Selected Papers from the Eleventh International Conference on the History of the Language Sciences*. Philadelphia: John Benjamins, 2011. 169–86.

Packham, Catherine. *Eighteenth-Century Vitalism: Bodies, Culture, Politics*. New York: Palgrave, 2012.

Partington, J.R. *A History of Chemistry*. 4 vols. New York: St. Martin's Press, 1961–70.

—— *A History of Greek Fire and Gunpowder*. 1960. Baltimore: Johns Hopkins University Press, 1999.

—— *A Short History of Chemistry*. 1937. New York: Dover, 1989.

Peterfreund, Stuart. "'Great Frosts and…Some Very Hot Summers': Strange Weather, the Last Letters and the Last Days in Gilbert White's *The Natural History of Selbourne*." *Romantic Science: The Literary Forms of Natural History*. Ed. Noah Heringman. New York: SUNY Press, 2003. 85–110.

Poirier, Jean-Pierre. *Lavoisier: Chemist, Biologist, Economist*. Trans. Rebecca Balinski. Philadelphia: University of Pennsylvania Press, 1996.

Porter, Roy. *Blood and Guts: A Short History of Medicine*. New York: Norton, 2003.

—— "Erasmus Darwin: Doctor of Evolution?" *History, Humanity and Evolution: Essays for John C. Greene*. Ed. James Moore. New York: Cambridge University Press, 1989. 39–70.

—— *The Greatest Benefit to Mankind: A Medical History of Humanity*. New York: Norton, 1998.

Redish, Angela. *Bimetallism: An Economic and Historical Analysis*. New York: Cambridge University Press, 2000.

Reill, Peter. *Vitalizing Nature in the Enlightenment*. Berkeley: University of California Press, 2005.

Robinson, Myron. "A History of the Electric Wind." *American Journal of Physics* 1962: 366–72.

Roe, Shirley. "The Life Sciences." *Eighteenth-Century Science*. Ed. Roy Porter. Vol. 4 of *The Cambridge History of Science*. New York: Cambridge University Press, 2003. 452–72.

Roger, Jacques. *The Life Sciences in Eighteenth-Century French Thought.* Trans. Robert Ellrich. Stanford: Stanford University Press, 1997.

Rose, S. Clifford. "An Historical Overview of British Neurology." Finger et al. 613–28.

Rousseau, G.S. *Nervous Acts: Essays on Literature, Culture and Sensibility*. New York: Palgrave, 2004.

Rowsell, M.C. *Ninon de L'Enclos and Her Century*. New York, 1910.

Russell, Michael. *The Chemistry of Fireworks.* 2nd ed. Cambridge: RSC, 2009.

Sargent, Thomas, and François Velde. *The Big Problem of Small Change.* Princeton: Princeton University Press, 2002.

Scales, Helen. *Spirals in Time: The Secret Life and Curious Afterlife of Seashells.* New York: Bloomsbury, 2015.

Scheurle, William. "A Reexamination of Coleridge's 'The Eolian Harp.'" *SEL* 15.4 (1975): 591–99.

Schiffer, Michael, with Kacy Hollenback and Carrie Bell. *Draw the Lightning Down: Benjamin Franklin and Electrical Technology in the Age of Enlightenment.* Berkeley: University of California Press, 2003.

Schofield, Robert. *The Enlightened Joseph Priestley: A Study of His Life and Work from 1773 to 1804.* University Park, PA: Pennsylvania State University Press, 2004.

——. *Mechanism and Materialism: British Natural Philosophy in an Age of Reason.* Princeton: Princeton University Press, 1970.

Scott, Arthur. "The Invention of the Balloon and the Birth of Modern Chemistry." *Scientific American* January 1984: 126–37.

Selgin, George. "Steam, Hot Air and Small Change: Matthew Boulton and the Reform of England's Coinage." *Economic History Review* 56.3 (2003): 478–509.

Shapin, Steven, and Simon Schaffer. *Leviathan and the Air-Pump: Hobbes, Boyle and the Experimental Life.* 1985. Princeton: Princeton University Press, 2011.

Shapiro, Alan. *Fits, Passions and Paroxysms: Physics, Methods and Chemistry and Newton's Theories of Colored Bodies and Fits of Easy Reflection.* New York: Cambridge University Press, 1993.

Shaw, Napier. *Manual of Meteorology*. Cambridge, 1919.

Sheehan, Jonathan, and Dror Wahrman. *Invisible Hands: Self-Organization and the Eighteenth Century.* Chicago: University of Chicago Press, 2015.

Siegfried, Robert. *From Elements to Atoms: A History of Chemical Composition*. Philadelphia: American Philosophical Society, 2002.

Simonyi, Károly. *A Cultural History of Physics.* Trans. David Kramer. Boca Raton, FL: CRC Press, 2012.

Smith, Christopher U.M. "Understanding the Nervous System in the 18th Century." Finger et al. 107–14.

Spencer, Colin. *The Heretic's Feast: A History of Vegetarianism*. Hanover, NH: University Press of New England, 1996.

Starr, G.A. "Defoe and China." *Eighteenth-Century Studies* 43.4 (2010): 435–54.

Stuart, Tristram. *Bloodless Revolution: A Cultural History of Vegetarianism from 1600 to Modern Times*. New York: Norton, 2007.

Thirsk, Joan, ed. *The Agrarian History of England and Wales,* vols. 5–6. New York: Cambridge University Press, 1967–2000.

Thompson, Carl. "Shipwreck and the Forging of the Commercial Nation: The 1786 Wreck of the *Halsewell*." *Shipwreck in Art and Literature: Images and Interpretations from Antiquity to the Present Day*. Ed. Carl Thompson. New York: Routledge, 2014. 92–111.

Thurston, *History of the Growth of the Steam-Engine*. New York, 1878.

Trower, Shelley. "Nerves, Vibrations and the Aeolian Harp." *Romanticism and Victorianism on the Net* 54 (20009).

Vickers, Neil. "Coleridge, Thomas Beddoes and Brunonian Medicine." *European Romantic Review* 8.1 (1997): 47–94.

Wade, Nicholas. *A Natural History of Vision*. Cambridge, MA: MIT Press, 1999.

Wagner, Francis. *Hungarian Contributions to World Civilization*. Center Square, PA: Alpha Publications, 1977.

Watson, F. *Stargazer: The Life and Times of the Telescope*. Cambridge, MA: Da Capo Press, 2005.

Weinbrot, Howard. *Eighteenth-Century Satire: Essays on Text and Context from Dryden to Peter Pindar*. 1988. New York: Cambridge University Press, 2006.

Weinert, Friedel. *The Scientist as Philosopher: Philosophical Consequences of Great Scientific Discoveries*. New York: Springer, 2004.

Werrett, Simon. "Explosive Affinities: Pyrotechnic Knowledge in Early Modern Europe." *Making Knowledge in Early Modern Europe: Practices, Objects and Texts, 1400–1800*. Ed. Pamela Smith and Benjamin Schmidt. Chicago: University of Chicago Press, 2007. 68–89.

——. *Fireworks: Pyrotechnic Arts and Sciences in European History*. Chicago: University of Chicago Press, 2010.

Whitaker, Harry, C.U.M. Smith and Stanley Finger, eds. *Brain, Mind and Medicine: Essays in Eighteenth-Century Neuroscience*. New York: Springer, 2007.

Wigoder, Geoffrey, Fred Skolnik and Shmuel Himelstein, eds. *The New Encyclopedia of Judaism*. New York: New York University Press, 2002.

Williams, Glyndwr. *Arctic Labyrinth: The Quest for the Northwest Passage*. Berkeley: University of California Press, 2010.

Wilson, Eric. *The Spiritual History of Ice: Romanticism, Science and the Imagination*. New York: Palgrave, 2003.

Yeazell, Ruth Bernard. *Fictions of Modesty: Women and Courtship in the English Novel*. Chicago: University of Chicago Press, 1991.

Young, Charles Frederick T. *Fires, Fire Engines and Fire Brigades*. London, 1866.

For further information, see www.thebotanicgarden.net

For Product Safety Concerns and Information please contact our EU representative GPSR@taylorandfrancis.com
Taylor & Francis Verlag GmbH, Kaufingerstraße 24, 80331 München, Germany

www.ingramcontent.com/pod-product-compliance
Lightning Source LLC
LaVergne TN
LVHW020641300926
849161LV00004B/769

* 9 7 8 1 1 3 8 6 9 1 6 0 5 *